The
Untold Story
of a
Coast

JOHN JACOB PUTHUR

DEDICATION

I dedicate this book to the loving memory of my father, Puthur Lonappan Jacob

CONTENTS

Contents

FOREWORD

Admiral Arun Prakash, PVSM, AVSM, VrC, VSM (Retd.)
Former Chief of Naval Staff, India

Since new ideas are often seen as unorthodox, or even heretical, the history of scientific development, worldwide, shows a close connection with the acceptance of heterodoxy, or non-conformist views, in a particular society. Fortunately India has a sound historical tradition in this respect, and as Nobel Laureate Amartya Sen says in *The Argumentative Indian*, "It can...be claimed that the simultaneous flourishing of different convictions and viewpoints in India has drawn substantially from the acceptance of heterodoxy and dialogue."

The book that you hold in your hands, written by Commander JJ Puthur (Retd.), a naval hydrographer, is a good example of heterodoxy, in a field that he has studied, passionately, for many years. It is possible that Puthur's attempt to question conventional wisdom may evoke scepticism and indignation in some quarters, or even fuel a degree of controversy. But then, which iconoclastic seeker of truth – Copernicus, Galileo or Darwin – did not bring, upon his head, righteous anger and resentment from those who wished to cling to the known and familiar? Only time and healthy scientific debate will validate or debunk Puthur's theories, but keeping alive the spirit of enquiry is by itself an endeavour worthy enough of his labours.

On 7 June 2000, about a dozen warships, including an aircraft-carrier and submarines, were berthed in the calm waters of the Mumbai Naval Dockyard's tidal basin, sheltered from wind and tide by the South Breakwater. At about 4 PM the personnel on duty were astonished to see the ships suddenly start to surge forward, bringing unbearable strain on the ropes tying them to steel bollards on the jetty. A few moments later, the movement was reversed, bringing strain on a different set of ropes. After a couple of such powerful cycles, ropes began to part and a few vessels were cast adrift, leading to turmoil in the unwary dockyard.

The waters eventually subsided, and order was restored in the dockyard a few hours later; after tugs had helped ships and submarines back into their berths. The mysterious surge had resulted from the sudden collapse of a unique drydock being constructed under water in the vicinity. The accident, sadly, cost the lives of some labourers, caused minor damage to a few ships, and set the drydock project back by many years. The mishap startled the Navy, but the inquiry that followed failed to answer the question uppermost in everyone's mind—what was the root cause of this accident?

Puthur endeavours to explain this accident on the basis of a new hypothesis expounded by him. His theory also claims to shed fresh light on a whole array of coastal and hydrological phenomenon, such as silting of harbours, erosion of the coast and formation of sandbars, spits and ridges, as well as on some serious environmental issues. One had, so far, accepted the existing theories as explanations for all these without demur; but the author's compelling arguments urge you to look at them afresh. For a maritime nation such as India, whose coast is dotted with 200 major and minor ports, these are not issues of mere academic interest. They deserve our close attention, because the well-being of our coastal populace, as well as the protection of naval bases, ports and harbours, against environmental threats has a crucial bearing on national security.

Let me dwell briefly on the central theme of the author's hypothesis.

Mariners, hydrologists and observers of the seas, have always believed that waves indicate movement of sea water in a particular direction (most often towards or away from the coast), and since waves are created by the wind, the direction of wind and waves is always coincident. Hence one grew up at sea, believing that any object which fell into the sea would, eventually, be carried ashore by the waves. The most important deduction that emerges from this set of beliefs is that "waves move sediment", and the location as well as design of harbours and marine works have, therefore, always been configured to minimize the impact of wave-borne siltation. Design validation for harbours and marine works is undertaken through scale models which are subjected to tidal and wave action, simulated by pumping water through them.

Puthur, on the other hand, argues that the movement of water, as manifested by wind-induced waves, is a complete illusion. Therefore all hydrological theories, as well as simulations, based on the assumption that siltation is caused by sediments carried inshore by seawater are flawed. Waves, according to him, denote only a vertical, up and down, movement of water particles, but seawater, actually, does not flow in any particular direction. Waves are, indeed, created by wind, not by physically pushing the water along, but by an aerodynamic phenomenon called Bernoulli's Effect.

His own complex theory of how sedimentation actually takes place is described, by the author, in lucid language, and with compelling step-by-step logic which comprises *The Untold Story of a Coast*. The reader is taken on a fascinating armchair travelogue of India's long coastline, laced with nuggets of the author's personal experiences. Puthur presents an unconventional view through a hydrographer's expert eye, of how nature's agents like wind, waves, tides and weather shape our seascape. Using historical and literary allusions, scientific reasoning and common-sense arguments, he provides the reader his own evaluation and critique of coastal and marine projects such as New Mangalore, Kochi and Vallarpadam ports, as well as the Sethusamudram Canal.

During the course of this informative voyage around India's coast, the author is at pains to introduce the reader to a great deal of esoteric but interesting nautical minutiae, necessary to obtain a full comprehension of his theory. Those hearing, for the first time, of terms such as the "Beaufort's Scale", "wave-rider buoys" or "littoral drift", need not reach for the dictionary (or Wikipedia) because Puthur provides clear and concise explanations. Having viewed much coastal activity through the prism of his theory of sedimentation, the author's inquiring mind then explores, in the last part of the book, the creation of the Indian coast after the splitting of the Gondwanaland super-continent 165 million years ago, and undertakes an investigation of monsoon related phenomenon.

Puthur is most passionate, and at his critical best and when he discusses naval projects – past, present and future. He has a great deal to say about the Naval Dockyard Mumbai, the Karanja Breakwater, the Karwar Naval Base and the South Jetty in Kochi. He brings up some disturbing environmental issues such as the possibility of mercury pollution at Karwar, and its impact on humans as well as ships. In some cases, such as the collapse of the new drydock in Mumbai, he claims to have predicted the impending disaster; and I fear that it is assertions of this nature, and some of his criticisms of long-standing theories and practices, which are likely to raise hackles in the Navy and elsewhere. The obvious questions that one can foresee are—"Why did he not speak up in time?" and "What is the point of raking up issues long past?" I have no intention of defending the author, because he provides adequate answers to these questions in the book, and it would be well worth the reader's time and effort to seek them out.

Puthur is a member of the Hydrographic or Survey Branch of the navy, a respected but self-effacing speciality embedded within a traditionally "silent Service". Of all the tasks that Surveyors undertake, perhaps none is as demanding and vital as creating or updating nautical charts of coastal waters. Survey ships spend months, unseen and unheard, in isolated locations, their crews working 12-14 hours a day, gathering and compiling underwater and topographic data. It is only when you examine a nautical chart at close quarters that you are struck by the enormity of effort, and the passion for accuracy that must have gone into its creation; a characteristic of the surveyor's ethos and outlook.

It is this commitment and personal dedication that, clearly, come through in the pages of this book. I have known Puthur since 1980, when he reported, as a Sub-Lieutenant, to the ship that I was commanding, to earn his Naval Watch-keeping Certificate. When I cast my mind back, two particular attributes of this young officer that particularly struck me then, were his composure under pressure, and readiness to speak his mind.

Although our paths crossed infrequently, during the intervening quarter-century, he seems to have retained these qualities. In an environment where forthright, dissenting or unorthodox views are, unfortunately, not always welcome, it is possible that Puthur may have often encountered rough weather during his career. But I am pleased to note that he has cheerfully and steadfastly stuck to his guns, and persevered in the best traditions of the Navy.

A few years ago, he happened to drop into my remote retirement home one day, and while reminiscing about old times, he shared with me, his views on many professional issues; lamenting the fact that they had generally been regarded as radical, and often met with disapproval in the

Service. I then suggested that since he had now retired, there was nothing stopping him from freely airing his strongly held beliefs, opinions and views; as long as they were not of a personal nature. I am glad that he took up my suggestion, because he now joins the select band of Indian Navy officers who have ventured to put pen to paper.

While offering his non-conformist views and theories, the author not only encourages his readers to be sceptical, but is prepared to be questioned and challenged about what he has propounded. This book would have served an important purpose if it enlightens, provokes controversy or evokes curiosity amongst its readers. Puthur, on his own part, feels that inquisitiveness is the key to unravelling and understanding the mysteries of the universe, of which many still remain beyond our ken.

Writing this book has been a labour of love for the author, and getting it published somewhat of a struggle. But I consider the resultant volume a most worthy endeavour because it has as much to offer the lay reader, as it does to the naval officer, maritime professional, hydrographer, coastal engineer, geologist, meteorologist, historian or archaeologist. It is my sincere hope that this book will find wide readership and a place on many bookshelves—in libraries, offices and personal collections.

ACKNOWLEDGMENTS

First of all, I thank God for having given me countless providential opportunities and insights to get to know the coast, and then enabled me to write this book with an overabundance of grace, strength and patience.

I did not write this book alone, but in a community, the community of hydrographers, mariners, coastal scientists and engineers, experts in dredging and reclamation, geologists, meteorologists, oceanographers and hydrologists. They provided substance to this book, some in person through personal chats, classroom lectures and presentations in seminars and symposia, and others through their books, theses and papers. It does not matter if I agreed with them or not. I am deeply indebted to every one of them.

I thank Admiral Arun Prakash, the former Chief of Naval Staff and my first Commanding Officer. He got me excited to do this book.

I thank Mr Ramankutty, my father's dear friend, for his constant support and encouragement as I struggled through the book.

I thank Shobha, my wife, Sharon and Sheryl, our daughters, for their incredible patience.

Commander John Jacob Puthur, Indian Navy (Retired)

WARNING

It is a great mystery that though the human heart longs for Truth in which alone it finds liberation and delight, the first reaction of human beings to Truth is one of hostility and fear. So the Spiritual Teachers of humanity, like Buddha and Jesus, created a device to circumvent the opposition of their listeners: the story. They knew that the most entrancing words a language holds are, "Once upon a time...", that it is common to oppose a truth but impossible to resist a story. Vyasa, the author of the Mahabharata, says that if you listen carefully to a story you will never be the same again. That is because the story will worm its way into your heart and break down barriers to the divine. Even if you read the stories in this book only for entertainment there is no guarantee that an occasional story will not slip through your defences and explode when you least expect it to. So you have been warned!

Reverend Father Anthony de Mello in 'The Prayer of the Frog'

(Reproduced with permission of publishers: Gujarat Sahitya Prakash, Anand, Gujarat, India)

"All People occasionally stumble across the truth, but most pick themselves up and continue as if nothing had happened."

—*Winston Churchill*

PROLOGUE

Indian Coast adorns the subcontinent like an exquisite necklace. This five thousand four hundred kilometres long necklace is ornamented with beautiful wave-lashed beaches, with sands white, brown, golden and glittering black. Sadly, many beaches are eroding. Some have disappeared. Ugly seawalls have taken their place. Future of many more is uncertain.

Besides the beaches, there are long stretches of verdant mudflats that ever remain sheltered from the waves. These coastal mudflats are home to countless species of flora and fauna that can survive nowhere else. Large tracts of these exclusive marine habitats are being turned into saltpans to meet the burgeoning demands of caustic-chlorine industry or reclaimed to build ports, airports, roads, factories, power plants and townships. That has put the survival of countless species at serious risk.

The coast is also studded with imposing headlands, majestic cliffs and sombre rocks that stand resolutely against the persistent lashing of the monsoon waves, year after year.

At both its ends are emerald-like adornments, lush green deltas, Indus Delta to the west and Ganges Delta to the east, which too are fragile ecosystems.

Besides what nature has grandly bestowed upon it, humans too have erected imposing structures. The most spectacular ones are the lighthouses, to help mariners find their whereabouts at sea and to steer clear of offshore dangers.

In addition, there are several majestic forts along the coast, but mostly in ruins. The former rulers of coastal provinces built these forts in eventually unsuccessful bids to protect their territories form being overrun by ship-borne invaders.

The most pervasive manmade features however are the ports and naval bases. Some of them are marvels of coastal engineering, but many, particularly the recent ones, face serious problems, not only within themselves, but also cause problems on the coast. It will be soon clear to you that the lack of understanding of the coast is at the root of all such problems, not only on the Indian Coast, but worldwide.

Though the story's primary focus is on the Indian Coast, the scientific inferences drawn from it are valid for any coast in the world, without exception. Every coast works the same way. Only the degree of dynamism differs, because the intensity of nature's elements that work on the coast differs, both temporally and spatially—winds and waves, tides and tidal stream, and rains and runoff.

The Indian Coast is however far more dynamic than most others in the world. The reason for that is the all too familiar, yet poorly understood phenomenon of monsoon, the Southwest Monsoon, also known as the Indian Monsoon.

The dynamism that the monsoon imparts to the Indian Coast also makes it highly vulnerable to human impact, which invariably comes in the guise of well-intentioned coastal and port development projects. But most such projects are based on myths, rather than on sound scientific principles. Some of the coastal myths even masquerade as scientific theories. Such spurious theories are then taught to the budding coastal scientists and engineers, not as something that may be disproved sometime or the other, but as unchanging laws of nature. It is they who eventually become the masterminds of future coastal and port development. That makes the future of the coast, not only Indian, but world over, even bleaker. The myths therefore need to be busted and the truth revealed. That is the aim of telling this story.

The Indian Coast came into existence nearly 165 million years ago, with the breakup of Gondwanaland. But here, the story begins at the Naval

Dockyard Mumbai, in these recent times, that is, about two decades ago. That was where I became curious about the coast. Incidentally, curiosity is the mother of science!

The story thereafter meanders through some of the recent happenings on the coast, some not so recent, and some yet to happen, but already on the anvils of planners and designers. The aim of examining these projects through an entirely different perspective is not to fault anyone, individual or institution, but only to understand how the coast actually works. That warrants occasional forays into the coastal science, which may be quite new even to the experts, but simple, so simple that one need not be a scientist to make sense of it.

Finally, after making sense of how the coast works, the story ends by telling the origin and evolution of Indian Coast, the 'Brief History of the Monsoon Coast'. But that comes after a brief telling of the story of Indian Monsoon, which has been, quite unobtrusively, shaping the coast all along, over the last 65 millions of years, that is, about 100 million years after the coast actually came into existence.

As you read through this story, you can track the coast on the Google Earth, occasionally zooming in to take a closer look. In the process, you too may stumble on the truth! Do pick yourself up and continue to explore the coast. But next time let that be on foot. That is the ideal way to get to know this marvel, the Indian Coast!

CHAPTER 1

MYSTERY OF THE SILTING DOCKYARD

Friday, 31 May 1991—I reported to the Naval Dockyard Bombay, on permanent duty. It was not yet Mumbai.

My appointment letter read '*for duties with Officer-in-Charge, ASD (B) Survey Unit*'. ASD (B) stands for Admiral Superintendent Dockyard (Bombay). He heads the dockyard. Even though the unit bears the name, it does not function directly under the ASD (B), but under the Commodore of Yard or C of Y, who heads the dockyard's Yard Services Department.

So I reported to the C of Y. He asked me to take over the duties of Officer-in-Charge. The incumbent had retired, prematurely, several months before. No one was appointed in his place. Thus I became the Officer-in-Charge, ASD (B) Survey Unit.

It was a gloomy morning, sky overcast with dark monsoon clouds, impatient to pour down. It did two days later. From the C of Y's office, I went in search of my new workplace in the sprawling dockyard. The Chief Recorder, the unit's senior most surveying recorder, whom I had known since long, was waiting for me in front of the unit's rather nondescript entrance. Otherwise I would have missed it. The small name board was barely readable. It was covered by a layer of dust raised by the passing vehicles. It was awaiting the monsoon rains for a wash down.

The unit was housed on the first floor of an ancient building. I learnt later that it was built in the year 1750, as the opium godown of old Bombay Harbour, at Ballard Bandar. Building across the road was the pepper godown. After courtesies, we climbed the creaky wooden staircase, dimly lit by a dust covered incandescent bulb, to the first floor. I had no reason to be elated, nor had any inkling of the long and arduous adventure with the coast that I was about to embark on.

The Chief Recorder led me to my new office, through the survey chart room. It was a small cubicle, with wood-framed asbestos panels, all painted pale green. There was no door. A green bunting that hung at the entrance, like a curtain, provided the privacy. An ancient wooden writing table dominated the small space. A green felt cloth covered the table, with a sheet of glass over it. Below the glass were bits of paper, with handwritten notes, the legacy of my predecessor. I had them removed and consigned them to the waste bin, without checking their contents, even before sitting down on the straight-backed wooden office chair, behind the table. There were two similar chairs in the front, for the visitors. A new green coir carpet covered the floor.

On one corner, a large wooden file cupboard stood. It held the unit's files since inception nearly a dozen years ago. The files held many interesting stories, like the pieces of a complex jigsaw puzzle, but with no picture on the box. Piecing together the puzzle into a coherent form soon became an obsession. That eventually took me on the long journey of adventure with the coast. The outcome is the story that you are reading, which remained untold until now.

An ancient two-blade fan whirled above, noisily. The traffic noise from the road below was often louder. As the days passed by, the puzzle gripped me. I heard little noise from either. The cubicle had two small windows, one facing the Wet Basin, to the south, and the other the pepper godown, to the west. For me, the two became my lookouts into nothingness beyond the cramped confines of my cubicle, whenever I pondered on the puzzle.

The Chief Recorder then introduced me to my team, who came in, one by one, into my cubicle. I knew most of them. Some were with the unit since long. My predecessor had left no 'brief' behind. So I had to learn the ropes on my own. That was not difficult. My team knew what had to be done. The unit's job was to manage the maintenance dredging in the dockyard.

What is maintenance dredging? Before I answer that, I must tell you what dredging is. For a simple definition—*dredging is underwater excavation of sediments and rock*. Definition is simple, but dredging is not. It is a complex operation that includes excavation, transport and disposal of underwater bed material. That needs some complex pieces of machinery mounted on a floating platform, together known as dredger.

Dredging to create a new coastal facility—port, dock or navigational channel—or to upgrade an existing one is known as capital dredging. It is usually a onetime affair. Maintenance dredging, as the name suggests, is to maintain an existing facility, to restore depths reduced by siltation. It is a continual activity, usually done at regular intervals. In the dockyard, it is annual.

That brings us to the term 'siltation'. The hydrographic dictionary defines it as the *deposition or accumulation of silt suspended in a body of water*[1]. Definition therefore assumes that silt can be suspended in a body of water. It is also not silt alone that gets deposited. Usually, it is a mixture of sediments— sand, silt and clay. When this mixture is rich in clay, which normally is the case, it is known as mud. Mud may be soft, medium or hard. In the dockyard, it is soft mud.

The dockyard silts every year during the monsoon, the season of rains. At Mumbai, it starts early-June, punctually, and goes on till end-August, sometimes even up to mid-September. Everyone believed that the dockyard silted steadily during the monsoon as soft mud suspended in water settled

---

[1] Entry 4747, Hydrographic Dictionary, IHO Special Publication No. 32 (Fifth Edition)

down. The view agreed with the definition. Anyway, by the end of monsoon, the depth in the dockyard went down by two to three metres. That made berthing and movement of warships and submarines difficult. The depth therefore had to be restored, quickly. That called for maintenance dredging.

The dockyard first experienced this siltation in 1979, soon after the completion of South Breakwater. See the Figure 1.1 below. First bout of siltation was severe. No one expected it. That necessitated urgent maintenance dredging, which however was beyond the scope of dockyard's large, but obsolete fleet of dredgers. So Navy called in the DCI, the Dredging Corporation of India.

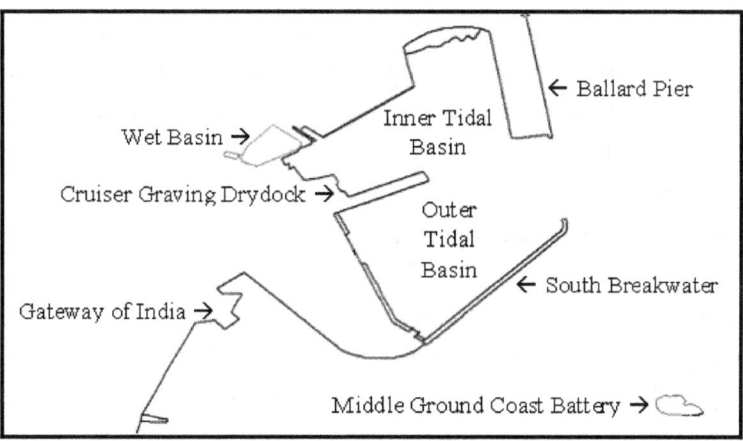

Figure 1.1: Naval Dockyard Mumbai

DCI deployed one of their modern dredgers, especially suited to dredge soft mud, the trailer suction dredger. In 1979, they dredged 1.2 million cubic metres of mud to restore the operating depths. It took them four months. The following year, during the monsoon, the dockyard again silted. Dredging became an annual routine, the post-monsoon maintenance dredging. The quantity dredged each year varied between 0.6 and 1.2 million cubic metres. No one knew why.

It is chaos in the dockyard during the maintenance dredging. Entire wharfs have to be vacated and left vacant for several days for dredging. Finding alternate berths for warships and submarines in an already congested dockyard is not simple. Vessels have to be frequently juggled around the available berths to meet various operational and maintenance demands. Movement of one vessel affects several others. That disrupts training and maintenance en masse. Siltation therefore costs the Navy lot more than the annual dredging bill, which alone may exceed ₹ 120 million now.

Almost everyone, including the experts, believed that the siltation was a natural phenomenon, hence unavoidable, something to be lived with. But it baffled me, on three counts! One, the quantity of mud dredged each year was disproportionate to the dockyard's size. How could the water in such a small dockyard hold so much mud in suspension? Two, the siltation was happening too fast. The dockyard was silting up in less than four months. How could so much suspended mud settle down so quickly from water that was flowing, almost continuously? Three, the siltation was only happening during the monsoon. Why was it so seasonal? If it was only a matter of suspended mud settling down, it must happen year round.

Soft mud, dredged from the dockyard, is about three times denser than the seawater. So the mud must naturally remain on the bed alone, held down by gravity. If the mud is to be held in suspension, water must flow swiftly, like the runoff rushing downslope during a heavy downpour. The runoff is sometimes muddy, because of the mud held in suspension.

Water flowing through the dockyard is tidal or tidal stream. As the tide rises, tidal stream flows into the dockyard to raise the water level. It is also known as the flood stream. As the tide falls, ebb stream flows out to sea bringing the level down. It takes about six hours for the tide to go up or down by two to three metres, therefore the tidal stream, either way, is slow, in fact, too slow to bear any mud in suspension.

Also, at any place on the coast, the tide seldom falls below a minimum level known as the Lowest Astronomical Tide or LAT. It occurs once in 19½ years, the tidal epoch. However, to serve as a reference level for measuring depths, the Chart Datum is established for each place on the coast. It is arbitrarily set a shade below the LAT for the place. It then remains valid for a long time, unless the sea level changes drastically, say, due to rapid global warming or cooling.

For the present, we can assume that the tide seldom falls below the Chart Datum, though it does at Mumbai, occasionally. The datum at Mumbai was set a shade above the LAT, perhaps inadvertently, and no one felt the need to change it thereafter. Ordinarily, the rise and fall of tides only happens above the datum. As a corollary, the tidal stream also must flow only above the datum. In other words, the tidal stream is a surface phenomenon. Since dockyard's bed is entirely below the datum, the tidal stream is incapable of moving mud even along its bed.

Yet, it is a widely accepted theory in coastal science that the tidal stream moves sediments, not only in suspension, but also along the bed. I shall tell you in the Chapter 4: Working with Models, how this theory may have come about. Anyway, at that time, I too thought it was true. I was taught that in my courses. Moreover, the water in the dockyard always appeared muddy, with a greenish-brown colour. In other words, the tidal stream did seem to hold mud in suspension.

Anyway, not with the intention to challenge any theory, but only to somehow estimate the rate of siltation in the dockyard, I decided to measure the amount of mud actually suspended in its waters. I had water samples collected from many places, from different depths, even from the bed, during flood and ebb, and also before and during the monsoon. I had measured volume of each sample filtered using a filter paper. But there was no mud in any of the samples, except those from the bed, which, in every case, was not water, but fluid-mud, a thick creamy solution of mud that did not easily pass through the filter paper. The dockyard's bed was covered by a layer of fluid-mud, but there was no mud in suspension, anywhere and anytime! The siltation became more intriguing.

The unit carried out regular surveys in the dockyard, mostly dredging related—pre, check and post dredging surveys. These surveys were done jointly with the representatives of DCI. Pre-dredging survey, as the name suggests, is done before the dredging, but only after the withdrawal of monsoon. It gives the extent of siltation during the previous monsoon.

Check-surveys are done to monitor the progress of dredging. Post-dredging survey is done after the completion. From pre and post-dredging surveys, the volume of mud dredged is computed, and DCI paid accordingly.

Besides the dredging surveys, the unit also surveyed the dockyard once every month soon after the completion of maintenance dredging. These monthly surveys went on till the onset of the next monsoon. But the surveys showed no change in depth. In other words, the dockyard did not silt during the non-monsoon months! So I discontinued the practice.

The unit therefore held enormous amount of survey data, but all of the non-monsoon period, when the dockyard did not silt. There was no data for the monsoon when siltation was actually taking place. Therefore, from the available data, it was impossible to deduce the rate or pattern of siltation. I needed fresh data for that.

So I decided to survey the dockyard during the monsoon. I suggested that to my team. They were not enthusiastic. Surveying in the rain was neither simple nor pleasant. Nevertheless, I went ahead with the survey, but not of the entire dockyard. I chose few representative lines, which I decided to survey regularly from the beginning to the end of monsoon. I felt that would be sufficient to indicate the rate of siltation and its general pattern.

I had the lines surveyed few days before the onset of monsoon. The dockyard still had the depth achieved during the last maintenance dredging. The rains started, as usual, in the first week of June. It rained heavily for about a week. During the brief lull that followed, I had the lines resurveyed. The data made little sense. The dockyard had silted fully, to the brim, so to say, and all in just a week of the monsoon! Siltation was happening much faster than I had imagined. It seemed as though the mud was being pumped into the dockyard at a tremendous rate.

A few days later, during the next lull, I had the lines surveyed again. There was no change in the depth! The dockyard stayed silted, fully, as before, no more and no less, as though someone had turned off the mud supply. Where was the mud coming from and how? Why did it stop flowing in after few days?

I talked to the experts. I read their books and papers. The mystery only got deeper. That got me curious about the coast, and at the same time, suspicious of the existing body of coastal knowledge. I soon realised that I was dealing with an enigma—puzzling, ambiguous and inexplicable. But with more than a century of research into the coast, it may even seem rather preposterous to suggest that there was anything enigmatic or even mysterious about it. After all, who does not know the coast?

Let me therefore test your coastal knowledge. I assume you have made, at least, one visit to the coast, a sandy beach, anywhere. Recall one of your pleasant beach experiences. What comes to your mind?

It may be the grand spectacle of sea, with a sunrise or sunset of exquisite splendour. Or it may be the endless roar of sea. Some find it enthralling, like listening to a philharmonic orchestra in concert. Or it may be the taste of salty sea air. Or you may recall the smell of sea, of the marine life, living or rotting. The most exciting part of any coastal visit would of course be playing with the breaking waves, of being tossed about, accompanied by joyous screaming and shouting.

Therefore, on the coast, we can hardly escape one or more of the sensory experiences, all arising from the sea. In fact, the sea literally swamps our senses. In other words, sea is a compelling multi-sensory experience. What about the coast?

For most of us the coast just happens to be a firm place, a terra firma, to experience the sea. In fact, most of people go to the beach only to be enthralled by sea and its waves. Is it therefore any wonder that we know the coast only in terms of the sea?

We know the coast only as the land adjoining sea, bank of the sea, border between land and sea or just plain seashore. That is how the dictionaries define coast. Even the dictionaries of geography, geology, hydrography and geomorphology define the coast no differently. But such a definition tells little about the coast or its working. So there is a need to coin a new definition. Before we can do that, we must know the coast well enough. That is the aim of telling this story—to know the coast and how it works.

Let me therefore begin this story with the premise that coast is an enigma, an unknown. That is a better way to start the learning process. Nevertheless, for a hydrographer to admit that the coast is an unknown may seem rather strange or even out of place, because it is his job to survey the sea and its adjoining coast. It is known as the hydrographic survey. So, before we begin unravelling the enigma the coast is, let us get to know how the hydrographer goes about his work, surveying the sea and coast.

CHAPTER 2

HYDROGRAPHER'S WORK

Hydrographic surveys are done mainly for mapping the seas or nautical charting. The primary purpose of nautical charting is to ensure safe marine navigation, for the seagoing vessels to sail the seas, safely.

The principal end product of a typical hydrographic survey is the navigational chart. The chart is the fundamental and indispensible instrument of marine navigation. No seagoing vessel can be legally put to sea without a valid set of up-to-date charts.

The chart however is not for general use, like a land map. It requires some training to be able to use it. Moreover, it has few familiar features to show, particularly, across the water spread that covers most of its space. Instead, it shows depths in figures, metres and decimetres. For example, the figure '11_8' on a chart represents a depth of 11.8 metres, exactly where it is shown.

Also on the chart, the water spread is shaded white, blue or olive green depending on depth. Deep navigable areas are in white. Areas shaded blue are shallow, which on a coastal chart is usually less than five metres deep. Intertidal zone, the zone between the low and high water lines, is shown in olive green. The land beyond is usually coloured yellow, with features indicated in brown or black symbols. Magenta is another common colour on the chart to indicate the navigational features and zones.

Besides depths, the chart shows every known or suspected offshore dangers at their best known positions, using appropriate symbols—shoals, rocks, islets, wrecks, buoys, beacons, fishing stakes and anything else that may impede the safe passage of a vessel. It also shows conspicuous features on the coast and land beyond. These include lighthouses, beacons, towers, spires, chimneys, buildings, hilltops and anything that may appear conspicuous to the mariners at sea. With the help of these charted onshore features, they can find their whereabouts at sea and identify the coast they are heading to or passing by.

In addition, the chart presents a host of other information to ensure safe marine navigation—tidal stream, tidal characteristics, nature of seabed and more.

The most important data for marine navigation is depth. Measuring depth therefore is the principal activity during the survey. It is called sounding, because an echosounder is used to measure depth. Echosounder transmits a beam of ultrasonic sound to the seabed and records the time taken for the echo to return. Then, from a pre-set speed of sound, it computes the depth.

The speed of sound through seawater varies with temperature and salinity. Therefore, to obtain accurate depth, the hydrographer also measures the speed of sound underwater periodically, and resets the echosounder, whenever the change is significant.

Depth varies from point to point. So when sounding, the hydrographer must fix the position of the sounding vessel, survey motorboat or ship, at regular intervals. In the past, position fixing systems were land based. Their onboard components derived positions at sea from known positions on land, where the onshore components were set up before commencing the survey.

Advances in space technology brought in satellite based position fixing systems. Latest in the genre is Global Positioning System or GPS. Today, it is a household name. It is fitted in cars and buses, and even on cell phones.

For hydrographic surveys, Differential Global Positioning System or DGPS is today the mainstay for position fixing. It is both accurate and easy to deploy. DGPS receiver receives data from an array of satellites, just like the GPS. In addition, it gets corrections, via a radio or satellite link, from the reference stations set up at accurate positions on land, usually on the lighthouses. The corrections remove some of the errors due to the aberrations in the orbits of GPS satellites and due to atmospheric conditions. That makes the DGPS far more reliable and accurate than the GPS.

Depths vary with time, due to the rise and fall of tides. Height of tide matters where depths are shallow, particularly near the coast. Therefore, when surveying in coastal waters, it is necessary to observe the tides. Observing tides means recording the height of tide against time continuously or at regular intervals. That is done using a tide pole or gauge, which is set up at a sheltered place within the survey area, usually on a jetty or wharf. The pole or gauge is then levelled to a benchmark, a mark set firmly on the ground, at a known height above the Chart Datum. Levelling connects the zero of pole or gauge to the datum. Heights of tide recorded are thus referenced to the datum.

Every sounded depth is then reduced to the datum by deducting from it the height of tide at the time of sounding that depth. Depths shown on the chart are therefore the least available, wherever shown, whatever the height of tide. To obtain the depth at any time, the mariner adds to it the predicted height of tide, which he derives from the tide tables. Tide tables for Indian Ports and select foreign ports are published annually by the Survey of India and also by the National Hydrographic Office, both headquartered at Dehradun.

The chart does not show every sounded depth, but only a select few at their positions, while ensuring that there are none shallower around those shown. That leaves enough blank space on the chart for the mariner to plot his vessel's track and record the details of time, course, speed and distance to go along the track.

Though the chart appears with plenty of blank space, during the survey, the entire water spread is sounded closely, leaving practically no blanks. But that may not be possible in shallow waters, particularly along the coast and near rocks and reefs, where the survey motor boat could run aground or otherwise wrecked. Such shallow areas are however not crucial to navigational safety, as long as they are marked clearly on the chart, even erring on the side of safety. Seagoing vessels normally keep to deep waters, hence well clear of the shallow areas along the coast. Areas along the coast therefore remain poorly surveyed.

But, at the end of the survey, there must remain no 'unknown' underwater dangers in the areas normally navigable to seagoing vessels. That is the desired outcome of the survey. Within such navigable areas, the hydrographer must find every significant feature on the seabed, fix its position and measure the least depth over it.

To make it doubly sure that there are no unknown underwater dangers within the navigable areas, it is now mandatory to sweep the seabed with a device known as side scan sonar. The sonar displays the image of the seabed showing all features that stand proud. In the past, when safe depth had to be accurately delineated, say, in a navigational channel, the hydrographer swept the channel bed mechanically with a weighted wire slung at that depth usually between two boats running parallel. Seldom was any feature on the bed shallower than the wire-depth missed. Wire sweep was very cumbersome. It is no longer in use. Echosounders and side scan sonars have become extremely accurate and reliable.

The hydrographer has other options to make the survey area absolutely safe for navigation. He can nowadays scan the seabed with a magnetometer to rule out the existence of any steel objects, for example, an inconspicuous mast of a buried wreck. Also, when in doubt, he can send down an underwater video camera or even a diver to check out the suspected feature on the seabed. That is the kind of care that goes into ensuring navigational safety. Environmental cost of a ship breaking her bottom by running into an unknown underwater danger amply justifies every effort.

During the survey, the hydrographer also measures the rate and direction of tidal stream, which nowadays is done using a currentmeter. In the past, it was done by tracking a drifting buoy with a flag tied to it. Tidal stream affects the vessels, both at anchor and underway near the coast, hence of interest to the mariners.

He also collects samples of the seabed sediments. He then analysis the samples and classifies the seabed as rocky, sandy or muddy. Nature of seabed is then shown on the chart using appropriate symbols. Mariners need to know that to choose where to anchor their vessels. Dropping anchor on a rocky seabed may shatter it. Soft mud may not hold the anchor well enough. The vessel may then drag her anchor and drift into danger, particularly when it gets windy.

He also collects a host of other environmental and oceanographic data, which also include data of marine flora and fauna. That is done mainly to further the understanding of the marine environment. That is about the work he does when it comes to surveying the sea.

When it comes to surveying the coast, he has comparatively little to do. He only has to delineate only two lines, the low and high water lines.

Low water line is the seaward limit of the coast. It is also known as the zero-sounding line or zero-depth contour. It is delineated during sounding, provided it is safe and practicable. It calls for careful planning and execution. Waves and nearshore rocks and dangers make it quite risky. The survey motor boat running aground or being washed ashore is common when surveying the low water line. Many hydrographers have courted trouble trying to 'catch' the low water line, which sometimes is seen as act of daredevilry in the field of hydrographic surveying. The low water line can also be fixed from coast, on foot. That too is no less risky and arduous.

Despite the attendant risks, the low water line has little to do with navigational safety. It is considered as the seaward limit of the nation's land territory, hence the baseline to measure the maritime boundaries—territorial sea, contiguous zone and exclusive economic zone. The charts are official instruments to show the maritime boundaries of the coastal state, hence the need to delineate the low water line.

On the other hand, delineating the high water line is quite simple. The high water line is the maximum extent the sea rides into land when the tide is highest. That happens during the springs, full and new moons. High water line is easily discernible on a cliffy coast or a steep beach. On a gently shelving beach, it is more difficult to discern. Usually, there is a distinct change in gradient at the high water line. That is also where the driftwood and flotsam or floating debris collects. Land vegetation, grasses and weeds, usually begins at the high water line. The high water line is the coastline that appears on the chart. On a densely vegetated mudflat, the seaward limit of the vegetation is taken as the coastline.

The hydrographer normally surveys the high water line on foot. It is known as coastlining. As he walks the high water line, he 'fixes' every bend or turn. And from the line, he remotely fixes the conspicuous features either side. Along muddy stretches or mudflats, where it may be quite slushy and so difficult to walk on, in the past, he used small flat-bottomed boat or rubber dinghy, but nowadays he flies over in a helicopter.

Therefore, almost everything known about the portion of earth's surface between the low and high water lines, which for all practical purposes is the coast, is based only on visual estimation from the high water line. No hydrographer actually surveys the coast. Actually, he cannot. He has no means at his disposal.

We can survey land easily, because we can move about easily on land. We can survey the sea, because we can sail it in boats or ships or even go under it in submarines or submersibles. On the other hand, coast has remained a difficult place to survey or to explore scientifically. It is too shallow to ply a boat, and uncomfortable to go about on foot, often dangerous too. There is no suitable platform currently available, manned or unmanned, to explore the coast, except an amphibious tank!

CHAPTER 3

SURVEYING WITH TANKS

Besides delineating the low and high water lines, Navy's hydrographers are sometimes called to measure the gradient of a coast, usually of a beach. It is known as the beach gradient survey. It is done on a stretch of beach, where an amphibious operation is planned. In the operation, the amphibious assault ships or landing ships, as they are normally called, land troops, trucks and tanks directly onto the beach. The operation is also called beaching. For safe beaching and pull out, the beach gradient must be known.

The survey however is quite rudimentary. Demand for accuracy is not as stringent as in the case of a regular hydrographic survey. But that does not mean it is easy. It can be an adventure. I had one in 1986. I was then the Executive Officer of INS Mithun, a survey craft based at Kochi. We were, at that time, stationed at the New Mangalore Port for a massive inter-services exercise code named 'Brasstacks'. The port was the staging area for the 'Blue' force. Surathkal Beach, north of the port, was the designated area for the landing ships to practice beaching, before they headed to the main exercise area, somewhere north. About half a kilometre of the beach had to be cleared for beaching, in other words, the beach gradient had to be measured. We got the orders late one evening.

On the following day, we came to the beach with the first light, well before the sunrise. A steady breeze was blowing onshore. Waves were already breaking on the beach. We chose a rubber dinghy for the survey. We had no choice in the matter. INS Mithun did not have a survey motor boat. Even if we had one, we would not have gone for it. On an open beach, where waves are breaking, the survey motor boat may not take long to be wrecked while attempting a beach gradient survey.

We had a portable echosounder onboard. It was too unwieldy for a dinghy. So we settled for a lead-line. Lead-line is a marked cord with a piece of lead at one end. To measure depth, the leadsman throws the lead-line ahead into water, easing the cord till the lead hits the bottom. Then he quickly takes off the slack as the boat moves forward. He reads off the mark on the surface of water as the cord gets vertical. That is the depth at that point. All that must be done quickly.

It is an ancient technique, but still finds use. Though laborious, it is quite accurate, but much depended on the leadsman's experience. Some of the old-time surveying recorders were experts with the lead-line. The depths they measured were as good as any digital echosounder.

For position fixing, we chose a sextant. It is a hand-held optical instrument to measure angles between marks erected ashore. Marks are usually large brightly coloured flags on tall bamboos erected at known positions ashore. In the past, before the electronic position fixing systems, sextants were the mainstay for position fixing.

Aboard a survey motor boat, two sextant-wielding anglers would measure two angles between a set of three marks, simultaneously. Immediately, they would shout out the angles, left first, then the right. Recorder notes the angles against a fix number in the sounding book. The hydrographer or sounder quickly plots the boat's position on a sounding board, after setting the angles on a three-armed protractor or station-pointer. He then tells the coxswain, one who steers the boat, to correct the course, if required. All that must be done quickly to ensure boat stays on the sounding line. That becomes quite a challenge when the sea turns rough. Most of the world's seas, close to the coast, were first surveyed by the early hydrographers using sextants and lead-lines.

Plotting was not possible in a dinghy, so we opted for a modified method—angle and transit method. In this method, the dinghy is steered on to a transit or two marks in a line, one behind the other, with a taller rear mark. Dinghy remains on the sounding line if the marks are in transit or seen as one. Where on this line the dinghy is depends on the cut-angle, usually the angle between the front transit mark and another at a known point. We had the cut-angles pre-worked for each point to be sounded on every sounding line. So, only one angler was needed.

As the dinghy nears the point to be sounded, the angler, who is continuously tracking the cut-angle with his sextant, would shout "standby." The leadsman would be ready with his lead-line neatly coiled in his hand, like a cowboy's lasso. The coxswain ensures the dinghy stays on transit, that is, right on the sounding line. On reaching the point, the angler would shout loudly—"fix." The leadsman would immediately fling the lead-line ahead into water. As the line straightens out, he would shout out the depth, say, "five metres". Recorder would record that depth against a fix number in the sounding book. By that time, the dinghy would already be on the way to the next point, without stopping. The angler would be tracking the next cut-angle. The leadsman would briskly coil the lead-line to be ready for the next dip. That was how we had planned to do the survey.

Use of sextants for position fixing is no longer in vogue. It is simpler and far more accurate to use a hand-held GPS, which today is readily and cheaply available. In 1986, there was no GPS. It came a year later. Sleek hand-held models came only recently.

We planned to sound 24 lines perpendicular to the beach, 20 m apart, each 200 m long. I selected the stretch after reconnaissance, having ensured that there were no 'charted' dangers in the approach.

We marked positions for transits for each sounding line using sextant and measuring tape. Fourth front transit mark from each sounding line, left or right, would be the cut-angle mark for that line. The shore-party would hold the marks or bamboos with flags at these points, for each sounding line, one after the other.

Sun was high up in the sky by the time we finished the marking. As the day got warmer, winds picked up. Sea got rougher. Waves got taller. We had a job to do. We rowed the dinghy to start sounding from the seaward end. Waves made it seem like white-water racing, dashing through the rapids of a mountain river. We soon got on to the first sounding line. In spite of the waves, we managed to stay on transit and take several dips. But as we neared the beach, waves took complete charge. The dinghy was tossed upside down. We found ourselves sprawled on the beach. Dinghy was floating away. Gear was strewn about. We were lucky to get away with only minor bruises.

The job had to be done. We quickly recovered the dinghy and gear, rowed to the start-point and began to sound again. But as we neared the beach, we met the same fate. Not to give up too easily, we attempted once more, hoping to get it right the third time. It was not to be. I had no intention of becoming the King Bruce of the 'spider' story. We sat on the Surathkal Beach soaked, wondering what to do next. The job had to be done, but how?

As the sun began to dry us out, we heard the rumble of tanks, the battle tanks. It was like a scene from a World War II movie. Several tanks emerged from their camouflaged hideouts behind the casuarinas and lined up on the beach, as if to launch attack on the 'cruel' sea, which by then had turned rougher. I guessed the tanks were taking part in Brasstacks, and were to embark the landing ships.

Then I saw two tanks heading out to sea. What a sight it was! The tanks, so massive and ungainly, were sailing like boats. They were amphibious tanks. After heading a short distance into sea, they returned to the beach. The waves did not matter to them. They were not tossed about like our dinghy. So why not use the tank for the survey?

I found the OC, Officer Commanding, in a nearby tent. Luck was on my side. He was my course-mate at the National Defence Academy, Khadakvasla. We were meeting after more than a decade. After catching up with the old times, I got down to business. Here was an opportunity to test the much touted inter-services co-operation that the academy was meant to foster. I asked him to loan us a tank for the survey. He readily agreed. But he told me that a tank could only make a short run of few hundred metres or so in water and had to cool down for some time before going out again. I told him of our need to survey 24 lines. He came up with a better solution. He agreed to line up, not one, but a dozen tanks, and more if we needed! He summoned the Subedar Major and gave the orders.

We got down to work. We had the tanks positioned at the spots we wanted, each forty metres apart. We made a run into sea, seated atop the first tank, sounding the first line, just as we would do in a dinghy. We did not have to row, nor were we tossed about. At the end of the line, the driver turned the tank around and lined it up on the next transit, twenty metres apart. We headed to the beach sounding. On getting back, we hopped off that tank, left it to cool, and got on to the next tank to resume sounding.

It worked so well that in about two hours we finished the job that otherwise would have taken us entire day, provided the sea was calm. We accomplished one of the finest beach gradient surveys ever. We also created history by surveying with amphibious tanks. It is quite unlikely that anyone would have used amphibious tanks for a survey, anywhere in the world.

If the coast was such a difficult place to survey, how do the coastal experts manage to study it? Simple, they do it on the model of the coast, a model complete with water, the hydraulic model.

CHAPTER 4

WORKING WITH MODELS

Hydraulic models are two types—wave and tidal. In the wave model, waves are simulated, and in the tidal model, rise and fall of tides. Sometimes the two are combined, the wave-cum-tidal model.

There is another classification based on the bed—mobile and rigid bed models. On the mobile-bed model, sediments or some material to simulate sediments are laid on its bed. It is used to study the actual movement of sediments underwater, usually in a canal, stream or river. After each run, the bed needs resetting. That makes it cumbersome.

Coastal studies are usually done on a rigid-bed model, where there are no sediments on the bed. The movement of sediments underwater is inferred from the movement of water on the model, which may be due to 'simulated' waves or tides.

It is easier and safer to study the coast on the model, but the outcome will depend on three factors. One, how accurately the model represents the coast and the adjoining seabed, the form, gradient and texture. Two, how accurately the model simulates the waves or tides. And three, how the waves or tides drive the coastal processes, provided they do.

The model, wave or tidal, is built from a large-scale navigational chart of the coast it represents. Large-scale charts are usually on scale 1:10,000 to 1:60,000. Scale 1:10,000 means 10,000 m on ground shows up as one metre on the chart. Bigger the ratio of length on the chart to ground, larger is the scale. 1:10,000 is therefore six times larger than 1:60,000.

Larger the scale, more are the details on chart. Also, each detail is represented with a better accuracy. But the area covered is small. As the scale gets smaller, the area covered increases, but the detail and accuracy decrease.

The hydraulic model is a chart in three-dimension, but on a very large scale, usually between 1:500 and 1:2000. Choice of the model's scale depends on the area to be covered, the detail to be represented and the room to house the model. A model on scale 1:1000 of a large harbour, like Mumbai, will require a hanger.

Unlike the chart, there is no relationship between the model's scale and the detail and accuracy. Though the model is on a much larger scale, the detail and accuracy are only as much as on the chart, no more and no less. Besides, the errors and inaccuracies on the chart are magnified on the model. No chart is without errors and inaccuracies. The hydraulic model therefore is not an improvement over the chart, neither in detail nor accuracy!

The chart represents the coast and adjoining seabed poorly, because those areas are very difficult to survey, sometimes impossible. But that does not matter to marine navigation, because vessels seldom venture so close to the coast. Marine navigation is the primary purpose of the chart.

Even without going into empirical evidence, we can say the form, gradient and texture of the coast and adjoining seabed affect the coastal processes. Such details are unnecessary for the mariners, hence the hydrographer does not spend much time and effort in collecting them, except in the passing, nor do the charts show any. Therefore, no matter how true the model may be to the chart, its coast and adjoining seabed may bear little resemblance to the real one. The coastal processes inferred from such a model therefore may be inaccurate, if not incorrect!

If the tidal model has the same scale along the horizontal and vertical, changes in height of tide may be imperceptible. For example, on a model with scale 1:1000, one metre change in height of tide on the coast will show up as a change in the water level by only one-thousandth of a metre or one millimetre. That is barely perceptible. So to make the tidal changes pronounced, the tidal model is vertically exaggerated, that is, the model's vertical scale is made larger than its horizontal scale. For example, if the model's horizontal scale is 1:1000, its vertical scale can be as large as 1:100 or larger. Extent of vertical exaggeration will depend on the tidal conditions on the coast. Where the tide range or the difference between heights of high and low tide is small, a larger vertical exaggeration may be necessary. The wave model does not normally need vertical exaggeration.

In a tidal model, water is released from a storage tank, through specially designed labyrinths or passages at a pre-determined rate to simulate the rising tide. Water from the model is then drained into a different tank, through another set of labyrinths, again at a pre-determined rate, to lower the tide. By regulating inflow and outflow through the labyrinths, the 'simulated' tide can be made to rise or fall, in quick succession, but proportional to the rate on the coast. Therefore, the height and time of tide on the model, after adjusting to vertical exaggeration and rates of inflow and outflow, will match that on the coast.

The vertical exaggeration also magnifies the rate of flow of the tidal stream. 'Simulated' tidal stream can be seen rushing through the model, like a swift mountain stream. But the rate may not match that on the coast, even after adjusting for vertical exaggeration and rates of inflow and outflow. The tidal stream on the model is gravity driven, that is, water flowing from higher to lower level. There is no other way to make it flow on the model. Therefore, depending on the bed gradient at different parts of the model, the rates of flow will differ.

That is not how the tidal stream flows at the coast. The seabed gradient does not affect it. It is a surface phenomenon, that is, the water flows only above the Chart Datum, whereas on the model, the entire water column flows from the surface down to the bed, no matter how deep, and at a speed dictated by the model's bed gradient.

Therefore, the popular theory that the tidal stream moves sediments in suspension and also on the bed has obviously come from studies on the tidal model only. That is not a coastal process!

The wave model requires wave data. That can be sourced from a meteorological office or collected by mooring a wave-rider buoy offshore. The buoy records the data in an onboard recording device based on its movement up and down with the waves. The data is then retrieved and processed. Say, they have the required wave data, processed and ready. How then do they simulate the waves on the model?

Waves are generated on the model by a motor-driven paddle. The motor that drives the paddle is controlled by a computer, into which the processed wave data is fed. The simulated waves are proportional to the input wave data, neat and regular, unlike the 'real' waves offshore.

When the paddle simulates waves, there is a matter that has been largely overlooked. The paddle on the model is no different from one used to propel a boat through water. What if the boat is tethered and paddled? The paddle's push would make water to flow along. That is the situation on the model. Therefore, besides simulating the waves, the paddle also makes the water to flow along. In other words, there is a measurable flow of water associated with waves simulated in a wave model. Is that so the case at sea?

Experts say yes. The water flowing in a wave model has given rise to much confusion about the coastal processes, particularly what the waves do at sea and on the coast. The story will deal more on that as we go along.

Not everything can be scaled down on the model. The two most important elements involved in every coastal process are water and sediments. These can hardly be scaled down. Sediments are denser than seawater. Coastal processes essentially involve movement of dense sediments through water. That is impossible to replicate on a model.

Therefore, on the model, wave or tidal, only floating particles are used to simulate sediments in suspension, even though the floating ones behave quite differently from the ones held in suspension. That has led to serious misunderstanding of the coastal processes, and in turn has affected the design of the coastal projects. See the Chapter 19: Bridging the Rubble-mound.

In the recent times, hydraulic models have given way to computer models. But the computer models are invariably based on the mathematics derived from studies on the hydraulic models, and then turned into software packages. Therefore, when it comes to inferring coastal processes, the computer model may not be an improvement over the hydraulic. It is only a digital replication of the processes observed on the hydraulic model.

The apparent advantage of a computer model is one of ease, but the outcome need not be correct. Yet it has found widespread acceptance among the coastal experts. Unless the mathematics that truly represents the coastal processes can be derived, a computer model, no matter how sophisticated, can only mislead. Therefore, we must first get to know the coast before we can model it.

Before we can get to know the coast better, we must first unravel the myths that surround it, particularly those that masquerade as scientific theories, though only as good as fairytales!

CHAPTER 5

COASTAL FAIRYTALES

The popular fairytale, *The Wild Swans*, by Hans Christian Andersen is a about a little girl Elise and her adventures. Once upon a time, a wicked witch cast an evil spell on Elise's eleven brothers and turned them into wild swans. She then sent them flying off to a distant unknown land. When Elise came to know, she set out in their search. After a long time and several adventures later, she met an old woman, who lived by a river. This woman taught her how to break the witch's spell and turn the wild swans back to her brothers. But for that, she must find them. So she set out along the river.

The following passage from the fairytale[2] should be of interest to us. It deals with Elise's experience on the coast, on the beach.

Elise said goodbye to the old woman and followed the river till she came to the place where it met the sea.

The great, endless ocean lay before her. But there was not a ship or boat to be

---

[2]Page 46, Fairytales of Hans Christian Andersen, Reader's Digest

seen—how could she go on?

She gazed at the countless pebbles on the beach, worn smooth by waves: glass, iron, stone, all had been shaped and subdued by the water, though this was softer, even softer than Elise's delicate hands. 'The waves just keep rolling,' she said, 'making the rough stones smooth. I will be just as tireless. Thank you for your lesson, you bright waves. Some day, my heart tells me, you shall carry me to my dear brothers.'

So on the beach, she saw 'countless' pebbles, which she thought were worn smooth by the waves, shaped and subdued by water that was softer than her delicate hands. Can you believe that? Well, if you have been witnessing the persistent lashing of waves on a beach, you may not doubt it. But we do not normally find 'countless' smooth pebbles on a beach, though there are exceptions.

Beaches at the foot of mountains are sometimes pebbly. For example, the Myrtos Beach in Greece, it only has smooth white pebbles. It is at the foot of two mountains, Agia Dynati and Kalon Oros. The pebbles, most probably, must have come from the mountain sides with runoff or a swift stream that rushed down long ago.

Elise followed the river to reach the beach. There is no mention of any mountain on the way. Therefore, she could not have seen 'countless' smooth pebbles on that beach, but for an odd one here and there. It is only a fairytale, so anything is possible. But as in this fairytale, many books on coastal geology and geomorphology claim the waves by their persistent hammering wear rocks into pebbles, then grind them into finer bits, sediments—gravel, sand, silt and clay.

Most people are under the impression that the waves wear down rocks much the same way the winds in a desert do. You may have seen pictures of grotesque desert rocks carved by the winds. But even in a desert, the winds cannot carve or wear down any rock, even the softest, no matter how hard they blow. What wears the rocks is what the winds carry as they blow over the rocks, the abrading sand. Desert winds can bear sand as they blow. Can the waves do that too?

Elise saw waves rolling in endlessly, as if coming from a distant land or sea. She imagined that they would go back too, some day. And when they do, they would carry her to her lost brothers, after all, there was no boat or ship in sight.

She is not the only one who thinks that the waves can move things from one place to another. Most people, after witnessing the endless 'arrivals' of waves on a beach, believe that the waves can move them, if not to some distant place, at least, bring them ashore. This belief is not limited to ordinary folk. It is even among the experts. But they do not suggest that the waves can move humans and other commodities from one place to another. Yet they would vouch that the waves do move sediments. That again is a popular theory in coastal science—waves move sediments!

During a storm, when big waves lash the coast, the sediments appear to have been moved, either on or off the coast. But the movement is apparent only after the storm abates.

Coast may have lost sediments to sea. That is erosion. Coast may have gained sediments and grown seaward. That is accretion. Coastal waters, inshore, within creeks or estuaries, or offshore may have become shallow by the deposit of sediments. That is siltation. Mouths of streams, rivers or creeks may have been blocked, partially or fully, by deposit of sediments. If the deposit is in the form of an underwater ridge that spans the mouth, it is a sandbar. If it is in the form of a visible ridge of sand that extends at an angle from only one end of the mouth, it is a spit. If a visible ridge of sand spans the mouth fully, completely blocking it, it is a sand-ridge.

Sometimes, after a severe storm crosses the coast, we get to see sand deposited inland that invariably stand much higher than the usual high tide level. The deposit may be either in the form of a long ridge or a small mound or hillock of sand. Ridge of sand along the coast is known as berm. The mounds or hillocks of sand on the coast are the dunes, coastal dunes.

The common denominator in all the above cases is the movement of sand, in a general sense, sediments. Before the movement had become apparent, the waves were lashing the coast. Therefore, it is quite reasonable to say the waves moved the sediments, more so, when there are no other entities apparently at work on the coast.

But that is only an assumption, because no one has observed the waves actually moving the sediments. Most probably, it may have been from this assumption alone, the theory that the waves move sediments took birth. Theories are invariably based on reasonableness rather than actual empirical evidence.

How do the waves move sediments? That is not something anyone can observe on the coast or at sea as it happens. That is impracticable. When the sea gets rough, as during a storm, it may be dangerous to remain on the coast or offshore, let alone observe how the waves are moving sediments. It is also not possible to simulate the movement of sediments that are ordinarily denser than water on a wave model.

The movement of sediments, purportedly by the waves, can however be deduced mathematically, most likely, by correlating the energy of waves with the size and density of sediment particles. A sediment particle can be moved from one place to another by application of certain amount of energy. Waves do display enormous energy. Sediments too have been moved. That provides the reasonableness. In other words, it is quite reasonable to say the waves move sediments.

The theory soon gained widespread acceptance among the coastal and marine sciences fraternity. So popular it became that it came to be taught not as a theory, which could be proved wrong, but as an unchangeable law of nature. No one thereafter questions it, because it is almost impossible to prove or disprove empirically, at least with the prevalent mindset. The theory then does not take long to become a dogma. Dogmas are seldom challenged! But this story will!

With that, let us go through another coastal experience. Unlike the Elise's on the beach, this one was on a rugged coast.

"We stand on a rugged coast and watch the waves strike blow after blow with the relentless persistence of a trip hammer. The display of vast power is impressive, and some disintegration of the rocky walls proceeds before our eyes."

These words are not from any fairytale. Sir Charles Lyell, the famous geologist, wrote them in his seminal work, the *Principles of Geology*[3]. No one

---

[3]The quote however is taken from Levin H. L., *Contemporary Physical Geology (Third Edition)*—page 491.

as yet has found anything wrong with the statement. Most probably, you too may not, particularly if you have witnessed the waves battering a rocky coast, with the 'relentless persistence of a trip hammer'.

Sir Charles Lyell's observation however goes to the heart of every misunderstanding about coastal processes. The coastal knowledge today has stemmed largely from visual experience, what we have observed from a distance or when the conditions were very calm. No one has dared to find out what may be happening on a rugged coast when the waves are lashing. That would be foolhardy to attempt. But the waves do not go on lashing indefinitely. After a while, the sea calms down. Even then, it may not be safe to take a closer look. We can only observe it from a distance. What do we get to see? Some loose rocks and boulders strewn about. Sometimes, we may also find some gravel or sand in crags and crevasses.

It is common knowledge that boulders, stones, pebbles, gravel, sand, silt and clay are created when rocks are broken down, somehow. These rock constituents vary only in size. Therefore, it is quite easy to assume that the waves broke the rocks down by their persistent hammering, more so, when there are no other agents apparently at work.

The notion that waves pulverise rocks into sediments comes from a snapshot view of the coast, usually from a single, brief visit. If you have been observing a rocky coast for months or even years, you would notice little change. Nothing usually changes on a rocky coast, even after the persistent lashing by waves for a long period. Most rocky coasts in the world have withstood intense waves, not only for centuries, but for millions of years.

The wave myths do not end here!

CHAPTER 6

MYTHICAL CURRENTS

The coast of Madras, nowadays known as Chennai, was a straight stretch of sand, a beach, albeit a narrow one, from the mouth of Adyar River in the south to the Ennore Creek in the north. The coast was featureless, but for the mouth of Coovam River that discharged about six kilometres north of the Adyar River. A kilometre and half north of Coovam, right at the seafront, stood Fort St. George, the seat of Madras Presidency. The fort is no longer on the seafront!

Those days, the ships calling at Madras anchored about half a kilometre east of George Town. Passengers and cargo were brought ashore in small boats known as masulas. Often, because of the surf or breaking waves, the masulas could not make landing. As the masulas neared the beach, coolies waded across and brought cargo ashore on their heads. Cargo loss was high. Hardy male passengers managed to wade across from the masulas. Those who could not, particularly the ladies in their Victorian fineries, were brought ashore on the heads of coolies, in round bamboo baskets. Sight was comical enough to inspire the artists of those times to paint the scenes.

Getting on or off the ships at Madras was indeed a cumbersome affair. That prompted the Madras Presidency to build a pier at St George Town, about a kilometre and half north of the fort. The 340 m long T-headed masonry Madras Pier was ready by 1876. But the waves made berthing almost impossible. Ships could be berthed only when the sea was calm, which was quite rare indeed. Therefore, to create a tranquil zone around the pier, they built a pair of breakwaters, one on each side, like embracing arms. The pier enclosed by the breakwaters came to be called the Madras Harbour. The entrance was from the east, through a gap between the breakwaters.

Even before the breakwaters, with only the pier in place, something strange had begun to happen on the coast. Sand began to accumulate south of the pier. The narrow beach south began to grow seaward or accrete. It got wider at the pier, tapering off south, like a long triangle of sand. The accretion continued after the breakwaters were built. Before long, north end of the accreting beach spanned the entire length of south breakwater. It was probably then known as the Marina Beach. But today, it is not a beach, but a shoreline fortified with a seawall! Chennai's Marina Beach is to the south of Coovam. Even that has an interesting story behind it. See the Chapter 15: Story of the Marina Beach.

As the beach south of the harbour began to accrete, the beach north began to erode. An accreting beach was acceptable, but not an eroding one. People had their homes and businesses near the beach, at George Town. The eroding beach was therefore hardened with a stone embankment or seawall in a bid to arrest the erosion. But as they built the seawall, the erosion spread north. They extended the seawall further north. Erosion too advanced north. Then they eventually extended it north up to the town limit, and gave up. Erosion continued north, unabated. In a matter of few decades, it created an indentation, an open bay, north of the seawall, on what was once a straight beach. The erosion has not ended yet!

The harbour's east-facing entrance began to silt as the sand migrated from the accreting beach south. On the other hand, along the seawall north of the harbour, erosion rendered the bed deeper. In 1904, the harbour engineers cut an opening through the north breakwater turning the harbour entrance northward, while sealing off the original eastern entrance. North side of the south breakwater was turned into 'South Quay', for berthing ships. The old pier was dismantled. The stretch where it stood became the West Quay. The enclosed area came to be called the Inner Harbour. Seawall north of the harbour was also turned into a wharf. That became the Outer Harbour.

Later, they excavated south into the accreted beach, at the western end of South Quay, to create the boat basin and timber pond. In early 1960s, they again excavated through the South Quay to create a large dock for berthing ships, the Jawahar Dock. Inner Harbour was renamed Dr Ambedkar Dock. Outer Harbour was enclosed by a set of breakwaters, the north and east breakwaters and the outer arm. It was named Bharathi Dock. See the Figure 6.1 below.

The approach channel extends north from the Bharathi Dock. It runs along the coast for about 2½ km before turning east to head into the deep waters. Accreted beach south, almost entirely, came to house the onshore facilities—godowns, stockyards, workshops, marshalling yards and more. That obscured the sea view of Fort St. George. That was how the major port of Chennai evolved, starting out as a pier in 1876.

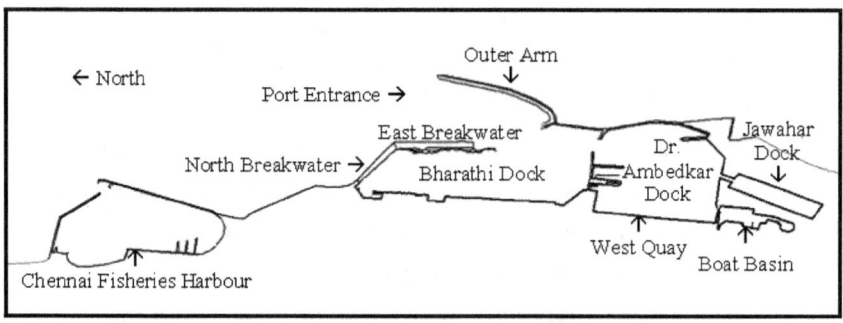

Figure 6.1: Chennai Port and Fisheries Harbour

The coast north of the port however continued to erode. The beach disappeared. Traditional fisherfolk had no place to park their catamarans. Years later, the Chennai Fisheries Harbour was built at the bay formed by erosion, just north of Chennai Port. See the Figure 6.1 above. Erosion now continues north of the fisheries harbour, unabated. Measures to control erosion have not worked.

The accretion south of the Madras Harbour and the erosion to the north led the experts to conclude that there was a natural drift of sand along the coast, from south to north. They called it the longshore drift. When the harbour impeded this drift, sand began to accumulate south, the upstream of longshore drift. No sand therefore went past the harbour north, the downstream of longshore drift. That led to starvation of sediments downstream, so the coast began to erode. All that seemed quite simple and logical too. But what powers the longshore drift?

Experts felt that a current of water flowing along the coast moved the sand in suspension. They called it the longshore current. They believed that the current was generated when waves came on to the coast at an angle. The angle decided the current's direction. For example, on the East Coast, during the Southwest Monsoon, the waves met the coast at an angle from the south, which according to them, sent the longshore current flowing north. During the Northeast Monsoon, the waves were from the north, again at an angle, hence the longshore current flowed south. Because the Southwest Monsoon waves were stronger, the northerly current was stronger. That in turn resulted in more sand being moved north. That explained the accretion south of the Madras Harbour and the erosion north.

The theory of longshore current was conceived only to explain the longshore drift. No one has yet measured this current on any coast in the world. That is impracticable, because of the waves and shallow depth along the coast. But on a wave model, because of the flow generated by the paddle, a current may appear to flow along the model coast, which probably was measurable too.

Nevertheless, longshore drift is real. It occurs on many coasts in the world, though not quite apparent until impeded by an obstruction—jetty, pier, breakwater or even an entire port. Experts have however found a way to measure the longshore drift. Since they have assumed that the waves were generating the longshore current, which in turn powered the longshore drift, they first measure the waves. That today can be done by mooring a wave-rider buoy offshore.

Then they measure the rate at which sediments is moved along the coast, purportedly by the longshore current. To do so, they dig a pit on the seabed offshore, but close to the coast. The pit is known as sand-trap. Then they measure the rate at which the trap fills up. That is easily done by sounding the trap at regular intervals. Here was a hitch, which no one took serious note of. On the upstream of obstruction, it was always sand that accumulated, whereas what filled the trap was mostly clay, with some silt and fine sand. Where did the clay come from?

The answer may hold the key to understand the coastal processes! That will be made clear a little later, in the Chapter 13: Coastal Sediment Dynamics. For the present, let us go along with the experts.

Then they deduce the rate of longshore drift from the rate at which the trap fills up. After several such sand-trap studies, in different wave conditions, they arrive at a statistical relationship between the longshore drift and wave data. Thereafter, using the relationship, they derive the longshore drift on any coast, by only measuring the waves. They no longer have to dig the sand-trap!

Let us now head to the West Coast, to a small coastal village by the name Chellanam, situated about ten kilometres south of Kochi Port. Sometime in early 1960s, the coast of Chellanam began to erode, suddenly, one monsoon.

By that time, the theory of longshore currents had come to be deemed a law, perhaps a dogma. Also, the experts believed the current was a norm on every coast in the world. Impeding it would lead to accretion upstream of the impediment and erosion downstream. That was the belief. But there was no obstruction anywhere along the Chellanam Coast that could have impeded the longshore current! The cause of Chellanam erosion therefore remained a mystery.

Waves came from sea, so the locals at Chellanam, who lost their homes and land to erosion, believed that the sea, for some unknown reason, had begun to attack the coast. They called it Kadalakramanam, meaning 'sea attack' in Malayalam. It soon became popular even among experts. They too were quite convinced that the waves were indeed attacking the coast. That was obvious from the solution they came up to counter the erosion—build seawall to fortify the coast against the 'attacking' waves.

Seawall however did not put an end to the attack. Before long, it crumbled. Everyone blamed the contractor. It was rebuilt under stringent supervision. Even that did not last. Now they have a massive seawall. Even that will not last long. The wave will take that too, albeit slowly. Nature is never in a hurry!

Erosion soon spread to other parts of Kerala Coast, and also to Karnataka Coast. Nowhere was any impediment to the longshore current. Therefore, it became a case of Kadalakramanam everywhere, though the local names perhaps differed. And the solution in every case was seawall.

But why was the coast eroding? Merely saying that the waves were attacking it did not seem quite scientific enough. There was a need for some theory to back it up. And there was one, the theory of littoral currents, by then quite well established and popular. The word littoral comes from the Latin word littoralis, with root in litus, meaning shore or coast. As per the theory, the waves near the coast generated currents capable of moving the sediments in suspension.

In spite of the advances in the instrumentation, no one has yet been able to observe the littoral currents, near any coast in the world, actually moving the sediments that are denser than the seawater, in suspension or otherwise. So it remained only a theory.

Without the backing of any field-data, some of the experts concluded that littoral currents caused the erosion at Chellanam and also elsewhere, whenever the sea turned rough. But why did the currents begin to erode the Chellanam Coast suddenly one monsoon? There was hardly a convincing answer.

Besides causing erosion, they also came to believe that the littoral currents moved sediments coastward, into the ports and harbours, to cause siltation. This belief cost the Indian Navy dearly!

CHAPTER 7

NAVAL DOCKYARD EXPANSION SCHEME

At the time of independence, Navy's fighting arm, the Indian Fleet, was based at Bombay. The fleet operated from the Naval Dockyard, where the facilities were quite primitive. Therefore, in the early 1950s, the Naval Dockyard Expansion Scheme or NDES was conceived to upgrade the dockyard.

Messrs Alexander Gibbs & Partners from UK were contracted to do the design. They were familiar with the Bombay Harbour. In 1914, they had designed the Alexandra Dock, which was later renamed Indira Dock. The Gibbs' team was soon ready with the design. Navy accepted the design.

There was stiff opposition to the NDES from the State of Bombay and Bombay Port Trust (BPT). The state was concerned with the aesthetics of Bombay's shoreline. They felt the NDES would obscure the historic landmark, the Gateway of India, and also mar the spectacular view of the harbour from the monument. BPT, on the other hand, felt NDES would hamper operations at Ballard Pier, their only passenger terminal. That was the era when people still travelled by ships to distant lands. Air travel was yet to become popular.

The matter was finally resolved in Navy's favour only after the then Prime Minister Jawaharlal Nehru personally intervened. Navy thereafter decided to execute the NDES in two stages. The work on the stage 1 began in 1954.

In the early 1950s, a central agency, the Central Water and Power Research Station, commonly known as the CWPRS, began to make foray into coastal and harbour engineering. CWPRS was set up in 1916 as a Special Irrigation Cell by the Bombay Presidency to deal with problems of irrigation and surface drainage. In 1925, the cell moved to a larger campus at Khadakvasla, Pune. The Central Government took over the cell in 1936. That was probably when it came to be called CWPRS. They were entrusted the task of planning and developing irrigation systems for the entire subcontinent. CWPRS now functions under the Ministry of Water Resources, Government of India.

Building hydraulic models was one of their principal expertises. They have been doing that since long, mainly to evolve and test the designs for irrigation canals, river training arrangements and overland drainage management. Their foray into coastal and harbour engineering was also through hydraulic modelling, to design new ports and to develop the existing ones. They sent some of their personnel abroad for training in coastal science and harbour engineering. With that CWPRS became a specialist institution in the field of coastal and harbour engineering.

In 1956-57, they built a tidal model of Bombay Harbour, at their Pune campus. BPT was at that time planning large scale developments at the port. So they needed the model.

The model was based on the navigational chart of Bombay Harbour, the Chart 2016. The model's horizontal scale was 1:1000 and vertical 1:100. The Naval Dockyard situated within the harbour also featured on the model.

CWPRS felt that the NDES would affect the hydrodynamics within the port and vice versa. So they sought to test the design on their model. Navy did not object. After testing, they felt the Gibbs' design would silt, and profusely. Navy would then incur huge and recurring expenditure on maintenance dredging, so they claimed.

The stage 1 was in progress as per the original design. It was probably too late to do anything about it. So they proposed redesigning the stage 2. Navy could not refuse. No one knew better. After more studies on their model, they came up with a new design for the stage 2. Navy accepted it. But what was wrong with the Gibbs' design?

When I was trying to unravel the mystery of the silting dockyard, I came to suspect the dockyard's design. From an old report, in one of the files in the wooden cupboard that stood in my small cubicle at the survey unit, I learnt that Messrs Alexander Gibbs & Partners had designed the dockyard in the early 1950s. There was no mention of CWPRS.

I felt the design document would probably throw some light on the siltation, provided the designers had foreseen it. I made a search for the document. No one in the dockyard apparently knew anything about it. The agency that executed the NDES by then had become the Director General Naval Projects Bombay or DGNP (B). They too could not locate the document. So I gave up the search.

Some weeks later, I had to meet the Staff Officer to ASD (B) for some personal work. As I was sipping a cup of tea in his office, I noticed on the wall a beautiful painting. It was the artist's impression of the NDES that Messrs Alexander Gibbs & Partners had originally prepared! But the dockyard on the painting hardly resembled the existing one. Only the Inner Tidal Basin matched. That was built as per the original design, the stage 1. So the stage 2 was changed. That eventually led me to CWPRS.

The Gibbs' design bore striking similarity to the adjoining Indira Dock, which they had designed in 1914. See the Figure 7.1 below. The Figure 7.2 below is a rudimentary reconstruction based on the artist's impression, superimposed on the plan of existing dockyard.

In the Gibbs' design, there was no South Breakwater, instead the Ballard Pier extended south up to the Middle Ground Coast Battery. The dockyard's entrance therefore was from the south. Indira Dock too has a southerly entrance. But that bothered the CWPRS. They believed that the littoral currents brought sediments from sea into the harbour. Southerly entrance faced the sea. The currents would therefore deliver sediments into the dockyard, so they believed. The dockyard would therefore silt.

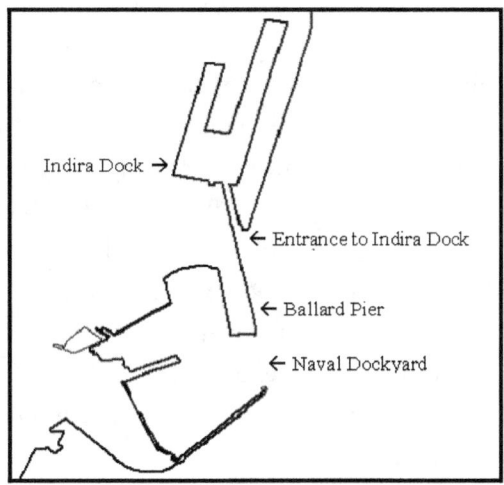

Figure 7.1: Indira Dock and Naval Dockyard

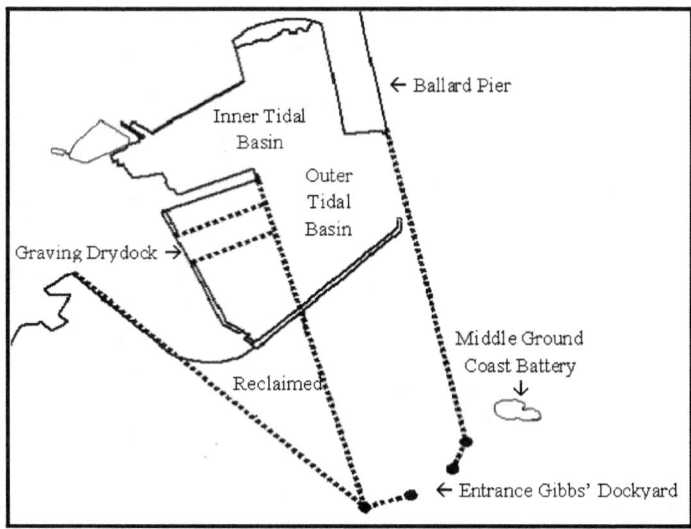

Figure 7.2: Gibbs' Dockyard Superimposed on the Existing Dockyard

They worked on a tidal model, which can tell nothing about the waves or wave-induced littoral currents. Therefore, based only on the theory, and with no other empirical evidence, they declared the Gibbs' design faulty.

Then they went about redesigning the NDES Stage 2. Their aim was clear and simple. Turn the dockyard's entrance away from the littoral currents, that is, away from the sea. They did that by introducing the South Breakwater. With that, the entrance came to be between Ballard Pier and South Breakwater, facing due northeast, certainly away from the sea. The littoral currents would no longer be able to deliver sediments into the dockyard. Hence, it would not silt, so they concluded.

Navy accepted the CWPRS design, without any scrutiny. Actually, they did not have any expertise to do so. They still do not have it. So the go ahead was given to execute the redesigned NDES Stage 2, most probably, without even checking back with Messrs Alexander Gibbs & Partners. What was the outcome of the design change?

CHAPTER 8

UNDERWATER DRYDOCK

The impact of design change was felt only in 1979, after the South Breakwater was completed. The dockyard silted. It continues to silt, year after year, every monsoon. Surely, the sediments did not come with the littoral currents, because the South Breakwater took care of that. Where then are the sediments coming from, and how? The answers would come later, in the Chapter 20: Mumbai's Mudflats. But for the present, let us examine the other impacts of the design change.

The design change led to the reduction in the area of Outer Tidal Basin and also of the adjoining reclaimed land. As a result, the wharfage and manoeuvring room in the Outer Tidal Basin reduced significantly. The reduction in the reclaimed area however affected the Navy more.

The Gibbs' design had envisaged a large graving drydock on the reclaimed area. See the Figure 7.2 above. The drydock was to come up on a natural underwater rock-outcrop, to give it a firm foundation. The rock-outcrop was therefore to be reclaimed. But the reduction in reclaimed area left it underwater, un-reclaimed, within the Outer Tidal Basin. Navy needed the drydock, but there was no room to fit one on the reclaimed area.

CWPRS therefore came up with a new design for a drydock to be built on the underwater rock-outcrop, without any reclamation. The design was unorthodox, most probably untried anywhere in the world, then and now. Yet the design was very simple. Essentially, it consisted of two walls jutting into the Outer Tidal Basin from the retaining wall of the adjoining reclaimed area, the Protective Retainer or PR Bund. The sill, where the lock-gate would sit, was at the other end. See the Figure 8.1 below.

They tested the design on their tidal model of Bombay Harbour, in 1978. After the testing, they concluded that the drydock would increase the siltation in the dockyard by about five per cent. No one took serious note of this rather small increase, because the redesigned dockyard was not expected to silt! Also, no one had any issues with the design.

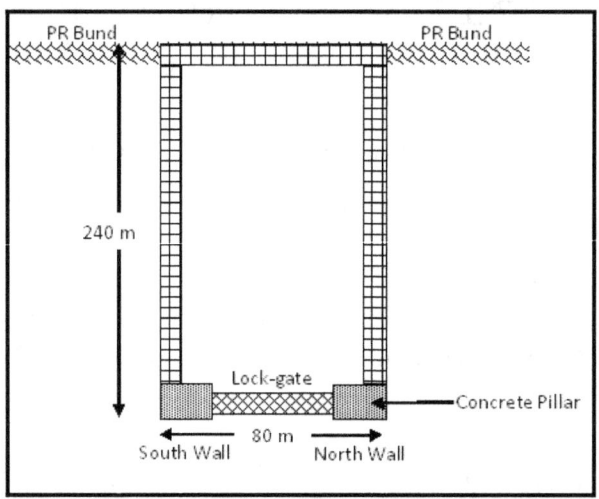

Figure 8.1: Design for the New Drydock

Navy did not however take up the construction in 1978, probably due to funding constraints, but did so about a decade later. The project was assigned to DGNP (B), who probably held the 1978 design. CWPRS had a relook at the design and found no cause to change it. The project was sanctioned. Capital dredging got underway in 1990.

The dredging essentially involved blasting the underwater rock-outcrop down to 9.75 m below the Chart Datum. The rock debris was recovered and stored nearby for later use in the construction. Only the mud was dumped at sea in hopper barges.

When I joined the dockyard, in 1991, the dredging was in progress. Two surveying recorders were attached with the project. They did the surveys jointly with the representatives of the dredging contractor. They used the contractor's survey boat and equipment. ASD (B) Survey Unit had no role.

The dredging was completed by end May 1993, barely few days before the onset of monsoon. At that time, the contractor's survey boat developed some trouble. They could not do the post-dredging survey, but were desperate to complete it before the rains. Without waiting to repair their boat, they sought my help. I sent a survey motor boat with crew. The survey confirmed the completion of dredging. On the bed of the Outer Tidal Basin, along the PR Bund was a rectangular pit, about 300 m long and 150 m wide. It was also uniformly deep—9.75 m below the Chart Datum.

The monsoon rains started soon after. As usual, the first spell was intense and lasted many days. During the lull that ensued, I ordered the re-survey of the dredged area. I knew what the result would be, but wanted to confirm. The entire area had silted back to the original bed level—4.5 m below the Chart Datum! The rectangular pit was no more. It had filled up, albeit with soft mud.

The contractor obviously knew that this was going to happen. They must have experienced it every monsoon during the capital dredging. No wonder they were in such a hurry to get the survey done. Had they missed the deadline, they would have had to wait for another four months and again dredge the mud to claim their dues!

As per the design, the sill would be 9.5 m below the Chart Datum. Therefore, siltation would pile up about five metres of mud in front of the lock-gate, and also in the drydock's approach.

Dredging near the lock-gate would be difficult, because the dredger cannot operate so close without the risk of damaging the lock-gate, with disastrous consequences. Therefore, the mud near the lock-gate would have to be carefully removed with a small grab dredger or by some other arrangement. A large dredger would however be needed to dredge the approach.

If the drydock needs opening during the monsoon, dredging would have to be done rapidly to counter the inflow of mud. By then it was quite clear that during monsoon the mud, literally, gushed into the dockyard. So to dredge the approach a large trailer suction dredger would be needed.

A year before, during the monsoon, the dockyard had to call in the DCI for dredging along the berth, where the aircraft carrier being undocked from the Cruiser Graving Drydock was to be berthed. DCI deployed one of their trailer suction dredgers operating at Bombay. The dredging was completed in two days, with the dredger working round the clock. The carrier was berthed there. But during the next pre-dredging survey, soon after the monsoon, we found the berth had fully silted. The carrier was all along sitting in mud. The berth had to be dredged again. Therefore, during the monsoon, every time the drydock needs opening, dredging near the lock-gate and in its approach would be inescapable. The earlier dredging would be to no avail. It would have silted back, almost immediately.

In addition, even after the dredging, as the lock-gate is being removed, for docking or undocking, we can expect huge amount of mud to flow into the drydock, literally, silting the dock floor. Clearing the mud would be both effortful and time consuming. Without that no work can be commenced, be it to lay the new dock-blocks or to clean the hulls of drydocked vessels. Incidentally, most drydockings are usually scheduled during the monsoon.

It would be impractical and also uneconomical to keep a large trailer suction dredger standing by for the entire duration of monsoon only to dredge the approach, whenever a docking or undocking is scheduled. A complex schedule would therefore have to be worked out to ensure the dredger's availability just before every docking or undocking. What if there is an emergency that warrants an urgent docking or undocking? Unless a dredger is operating at Mumbai, it may be several days before one can be made available.

I reported the matter. Soon I found myself explaining it to the ASD (B) in person, who then took me along for the drydock's high-level steering committee meeting chaired by the Flag Officer Commanding-in-Chief, Western Naval Command. I briefed the committee about the siltation. The matter was noted with concern. There were however some who were clearly displeased.

Few days later, I got a terse telephone call from a 'colonel', who refused to disclose his name, warning me to stay clear of the project, lest I burn my fingers. I knew what that meant. Few weeks later, I was tipped off for a transfer. Before long, I received the appointment letter. I was heading to Goa, to the Hydrographic School as an instructor. I welcomed that. Not that I had any choice.

But for a long time after the completion of dredging, the construction did not commence. I wondered if they had taken my suggestions seriously. But that was not the reason.

The delay was in the selection of the construction method. Originally, it was planned to build the drydock inside a cofferdam. Cofferdam is a temporary watertight enclosure, to build an underwater structure in a dry state, such as foundation piles on the riverbed or seabed. It is dismantled after the construction.

Building a cofferdam, large enough to house a huge drydock, proved not only difficult, but also impracticable. It would occupy almost the entire space in the Outer Tidal Basin. That would severely hamper berthing and movement of warships and submarines. The dockyard was already quite congested. The drydock would take several years to build. Only after that can the cofferdam be dismantled. Navy did not want to get into such a tight situation and for so long. Alternate methods were explored. Finally, it was decided to build the drydock without a cofferdam! That meant building it underwater, by lowering blocks of concrete, one over the other, like a mason building a wall. The construction strategy was more bizarre than the design!

Because it was going to happen underwater, there was no question of the use of any mortar to bind the blocks. When the drydock would be dewatered after the construction, the hydrostatic pressure outside would be enormous. How would it withstand the pressure? Here came the ingenuity of designers.

They felt the friction between the blocks would hold the structure together, provided the blocks were sufficiently large and heavy. They worked out the friction necessary to make the structure strong enough to withstand the hydrostatic pressure. Then they worked out the shape, size and weight of the blocks to achieve that friction. Each block was therefore a solid 17-tonne box of pre-stressed steel reinforced concrete—3 m long, 2½ m wide and 1 m high.

As per their plan, to get the required strength, each wall would have to be 11 m thick, in two sections of equal thickness, outer and inner. Each section would again have two layers of concrete blocks, with a small gap between, which probably would be filled with rock debris and concrete.

The construction would take place in four stages. In the first stage, two concrete pillars would be built, 240 m away from the PR Bund and 40 m apart from each other, the drydock's mouth. The sill would be between the pillars. In the second stage, the outer section of each wall would be built, between each pillar and PR Bund, by lowering the blocks, one over the other, using a barge-mounted crane.

After the completion of outer sections, in the third stage, the lock-gate would be set on the sill and the drydock, or whatever it can be called, would be dewatered. Thereafter, in the fourth stage, the inner sections would be built in a dry state, with the outer sections sealed off by the lock-gate serving as a permanent cofferdam. That was the construction plan.

The pillars and sill were sturdily built inside separate cofferdams. Then the concrete blocks were lowered, one by one, onto a levelled bed, to build the outer sections from bottom upwards. No matter how carefully lowered, the blocks did not go straight down, but turned slightly, one way or the other, due to underwater currents or twists in the crane's wires or movement of the crane itself. The blocks were therefore misaligned, sitting unevenly against each other and over those below.

The blocks were too heavy for one diver or even a team of divers to push and align. Moreover, it was extremely dark underwater. The divers later checked the seating of blocks by groping in utter darkness. They found gaps between the blocks. To confirm, the contractor sent down a special underwater camera that could literally 'see' through the mud, similar to the ultrasound used to scan the human body. The divers were right. Blocks were not sitting properly. There were gaps almost everywhere!

It would therefore be impossible to dewater the drydock. I could not think of a way to plug so many gaps underwater. So I concluded the dewatering would not happen. Without dewatering, further construction was impossible. So that was the end of the drydock project, I presumed, hence took no further interest in it.

CHAPTER 9

WALL COLLAPSED IN DOCKYARD

I came on permanent transfer to Bangalore in April 2000. My appointment letter read 'for duties with the Officer-in-Charge, Naval Detachment Bangalore'. On reporting, I was assigned duties of Mess Secretary, Naval Officers' Mess Bangalore.

On 25 May 2000, around ten in the morning, one of the senior officers from the DGNP (B) walked into my office. He was associated with the drydock project since long. I knew him from the Mumbai days. He was on a private visit to Bangalore. I had put him up in one of the VIP suites. After the courtesies, our conversation, as expected, shifted to the drydock. He was aware of my reservations about the project.

He said with a sense of achievement, "Puthur, in spite of your reservations, we have completed the drydock. It's ready."

I was surprised. It was a while since I had heard anything about the drydock project.

I replied, "Sir, the drydock can be considered ready only after it has been dewatered. I doubt if that's possible."

He replied, "You're wrong. The drydock will soon be dewatered, in just a matter of days, in fact, as soon I am back at Mumbai. Everything is ready, and in place. We have completed all checks."

"What about the leaks?" I asked.

"Oh, those have been plugged and for good. There are no chances of leaks anymore." So saying, he got up to leave.

He had an appointment with his lawyer. I kept wondering how they could have plugged the leaks. Had they really done so, it meant trouble, big trouble.

As he was leaving, I told him, "Sir, if you are serious about the dewatering, then I suggest, you shouldn't be standing anywhere near the drydock, when it is happening. I think you should be somewhere else, where you can take some photos or film the event on a video. I would like to see it."

He smiled and said, "I'll keep that in mind," and left.

The Officer-in-Charge, Naval Detachment Bangalore, who was present during the conversation, asked me soon after, "You mean something will go wrong when they dewater the drydock?"

I told him, calmly, "It will collapse."

I did not elaborate. At that time, I wished I was wrong. I was in a state of disbelief. There was little I could do to stop the dewatering.

The date set for dewatering was 7 June 2000. Early that morning, lock-gate was set on the sill, shutting the drydock's mouth. The dewatering commenced soon after. As the water level dropped to one-third, late that afternoon, the north wall caved in, with hardly a whimper. Only the concrete pillar was left standing. The lock-gate was blown away.

I got the news following morning. A friend called from Karwar. He handled admin matters for the Project Seabird, the project to create the third naval base at Karwar. He called me, because, not so long ago, I had mentioned to him about the likely fate of the drydock, if they somehow managed to dewater it. I was not surprised at the news. But he was at Karwar. I wondered how he got the news so quickly. One of the colonels with the project was being deputed for the Board of Inquiry.

Several workers at the site lost their lives. I had not thought of that possibility. Official death toll was eight. Some of Mumbai's newspapers reported that. One paper ran a small story, with a benign title, 'Wall Collapsed in Dockyard'. Walls or even entire buildings collapsing during the monsoon were routine matters in Mumbai. I wonder if anyone paid any attention to the news. What went unreported was the wall that collapsed was nearly six metres thick and had cost billions of rupees to build, besides adversely affecting the Navy, for many years to come. Bangalore's papers were silent about the incident. It was unlikely the matter was reported anywhere else, but Mumbai. There was little remorse over the lives lost, because the dead were illegal immigrants from a neighbouring country— cheap construction labour.

As the wall collapsed, water surged in, causing strong eddies within the dockyard. Warships and submarines swayed about at their berths. Their gangways fell into water. Power cables tore off from the junction boxes, plunging the vessels into darkness. Berthing hawsers parted. Several vessels were set adrift. Everyone was taken by surprise. The dockyard was in chaos for a long time.

Why did the wall collapse? Simple, it did not withstand the hydrostatic pressure outside. Did the designers go wrong in their friction computations? Before I answer that, I must tell you how they plugged the leaks, which I thought was not quite possible. It took me a while to get to know that, but got it from the horse's mouth, so to say.

Few weeks after the event, one of the directors of the firm that built the drydock came to the mess, with a friend, to meet the Officer-in-Charge. I was introduced to him. When the Officer-in-Charge mentioned to him that I had foreseen the collapse, he was keen to know why it happened. According to him, they had taken every care to build it, that too, under the strict supervision of the designers. But I was keen to know how they plugged the leaks.

That, according to him, was simple. They did it by pumping in 'rapid-setting' concrete into the gap between the two layers of blocks. The concrete set almost immediately, within seconds of pumping in. To add to the strength, before pumping in the concrete, they had lowered several steel

54

wires with anchors into the gap. The concrete flowed into all the gaps between the blocks sealing them shut, thus making the walls watertight. But the concrete filling, even with anchor wires, offered little extra strength to the walls. No one expected that anyway. The walls were meant to withstand the hydrostatic pressure by the friction between blocks.

When calculating the friction, the designers, most probably, accounted for only seawater coming between the blocks, and nothing else, in other words, only friction between wet blocks of concrete. Soft mud is a ubiquitous feature of the dockyard's bed. Every diver knows that. Every hydrographer is familiar with it. There is no reason why the designers not have known it. Analysis of the bed samples precedes every coastal project.

The bottom layers sat entirely in the mud. Incidentally, that was where the hydrostatic pressure maximum. After the first layer of blocks was laid, a layer of mud soon settled on it. On that mud sat the next layer of blocks. The mud is slippery, almost grease like. One need not be an expert to know that with a layer of the grease-like mud between the blocks the friction would go down substantially.

Therefore, during the dewatering, as the water level dropped to one third, the hydrostatic pressure outside exceeded the 'reduced' friction between the blocks. North wall collapsed. There is no other esoteric reason for the collapse.

Navy needs a drydock at Mumbai. Rebuilding the collapsed drydock, without the fear of another collapse is quite simple, but doing so without tackling the siltation may be an exercise in futility. But how can we tackle the siltation? Before we can answer that, we must know why the dockyard silts. But to get there, we must know how the coast works.

CHAPTER 10

MAKING WAVE SENSE

Before we can go into the actual working of coast, we must understand the phenomenon called waves, which almost everyone believes to be behind whatever happens on the coast, directly or indirectly. But much of our commonsense about the waves at sea may not make the 'real' wave sense.

Most of us may have one or the other preconceptions about the waves at sea. The preconceptions, no matter how seemingly logical or scientific, are like wearing blinkers. They allow us to see only what we expect to see, while blocking out all counterevidence. But letting them go is also not easy. One way to circumvent them would be to get down to the very basics, by asking questions that a child would normally ask. Children are surprisingly free of preconceptions, until they are taught.

So let us begin by asking the most fundamental question—what is a wave? In a very general sense, it is an oscillation on the surface of water, anywhere, in a hydraulic model, pond or lake, or at sea.

What creates the oscillation or wave? It is created when the surface of water is somehow disturbed, for example, by tossing a pebble into a still pond.

Why does the water surface oscillate or becomes wavy when disturbed? Though not amenable to simple visualisation, wave forming on the surface of water is a simple process. It involves two factors.

One factor is the property of water. Water is incompressible, which means water, or for that matter any other liquid, can neither be compressed nor stretched by the application of physical force. That makes the hydraulic brake in a car work.

The other factor is gravity, the force that pulls everything towards the earth. It is also common knowledge that gravity keeps water level in any container, manmade or natural, as long as nothing disturbs it.

Let us now examine what happens when this incompressible water keeping level due to gravity is physically disturbed by an external agent. Let us say, this external agent, whatever that may be, briefly pulls up a small portion of the surface of water, somehow, to form a hump. Instead of hump, let us call it crest, which is a more familiar term associated with the waves.

Because water is incompressible, all around the crest, the level immediately and correspondingly dips to form a depression or trough. The trough will invariably be concentric to the crest. That is the first wave, one crest and one trough. If we take out a thin slice of the circular crest and trough, we get the familiar shape of a wave as shown in the Figure 10.1 below.

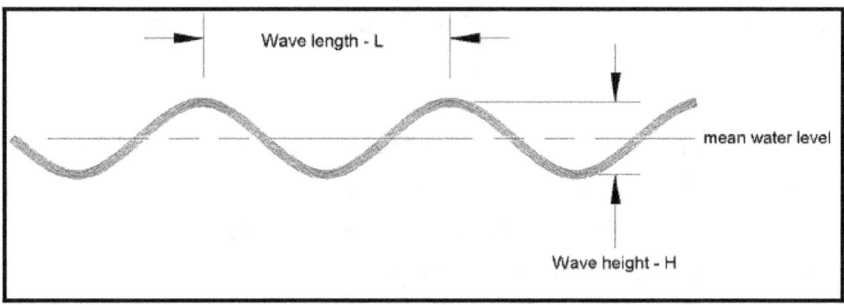

Figure 10.1: A Simple Wave

Immediately beyond, concentric to the trough, another crest forms, once again due to incompressibility. Thus the initial disturbance progresses outward as expanding circles of alternating troughs and crests till the edge of whatever is holding the water is reached—hydraulic model's rim, pond's edge, lake's shore or the coast. That is one part of the process.

By incompressibility alone, the wavy surface thus created would have remained, until something pushes down the initial crest to the original level. That is where gravity comes in. As soon as the external agent disturbs the water surface, even as the first wave is being created due to incompressibility, gravity gets to work to restore the level, literally, by pushing down the crests and pulling up the troughs. But this gravity-driven level restoring mechanism has no brake system. It does not stop when the level is reached, but invariably overshoots. As a result, crests become troughs, and troughs become crests. That goes on. The action of gravity also spreads outward as waves, much the same way as the action of incompressibility. But the two processes work against each other, steadily petering out the waves to restore the level. Gravity eventually wins.

A pebble tossed into a still pond disturbs the surface. That makes ripples or small waves. Likewise, a ship or boat making way through the water disturbs the surface to make waves. Similarly, the paddle in a wave model also disturbs the surface by treading it to simulate the waves. Occasionally, a powerful undersea earthquake may suddenly push up or down the seabed. That in turn may disturb the sea surface above to make sometimes massive waves, which have come to be called the tsunamis. But most commonly, it is the wind that makes waves at sea.

Wind blows along the sea surface, just about skimming along. Therefore, by no stretch of imagination, does it tread the water, like the paddle. But the wave experts, world over, believe that the wind does somehow exert a pressure on the sea surface along the direction it blows. They call it the wind pressure. It differs from the atmospheric pressure, which acts perpendicular to the sea surface, from above.

Some of them believe that it may be the wind pressure that disturbs the sea surface to make the waves. But many are not quite sure, because that cannot be empirically confirmed at sea. Anyway, the following lines taken from one of the textbooks in harbour and coastal engineering do throw some light on the lack of understanding of the waves at sea.

"Many investigators (like Helmholtz, Jeffrey, Phillips, Miles, Hasselmann) in the past have attempted to explain the process of wind energy transfer through factors like pressure gradient across the wake (Figure 3.2), resonance of turbulent eddies in the atmosphere (Figure 3.3), shear forces based on logarithmic wind profile (Figure 3.4) and resonant interactions between different wave components. However the exact nature of the process of wave generation still eludes the scientists owing to its complexity. (Kinsman (1965)). The consequent formation and growth of waves is influenced by the wind pressure, its speed, fetch (the distance over which the wind, blowing over the sea surface, remains the same) and wind duration (the time over which the storm prevails) together with depth of water at the site."[4]

Nevertheless, the experts are quite convinced that there is something called the wind pressure that the wind exerts on the sea surface horizontally. They believe that this wind pressure makes the water on the sea surface to flow along in the direction the wind is blowing. That is used to explain some of the ocean currents. Apparently, lot of field research effort has gone into it. Meteorologists and oceanographers also believe in the existence such wind generated currents at sea.

When strong winds blow onshore, say, during a storm, there is an observable rise in the sea level at the coast. It is known as storm surge. The experts believe that the storm surge is caused by the wind pressure heaping water along the coast, because water can flow no further inland.

Wind does push a floating body in the direction it blows, provided it stands slightly above the sea surface. If the wind also pushes the water along, it can lead to some practical difficulties, like sailing a sailboat into wind. The sailboat cannot counter both wind and water flowing on account of wind, and yet continue to make headway using the wind. But it is possible to sail a sailboat into wind, though not absolutely head on.

---

[4] Chapter 3 on 'Ocean Waves', Harbour & Coastal Engineering [Indian Scenario], Volume I, National Institute of Ocean Technology, Chennai—page 182.

There is a familiar law in physics known as the Bernoulli's Law. It is a law, because it has been empirically proved, beyond any shred of doubt, a law as certain as gravity. You may have studied it in the school-level science as the principle behind the working of household flit-gun or carburettor in the car.

As per that familiar version of the law, when air, or any fluid, is passed through a narrow vent, its speed increases with a corresponding drop in pressure.

In a carburettor, the drop in pressure sucks up petrol from a small tank into the vent. The increased speed of air within the vent atomises the petrol, in other words, turns the petrol into fine droplets that blend with air. The mixture of petrol and air is then delivered into the combustion chamber through an inlet valve. A tiny spark from a spark plug ignites the mixture. That powers the car. But that is only a limited view of the law.

The Bernoulli's Law is essentially a statement of *the conservation of energy of a fluid*. It states that the total energy of a fluid stays constant, whether still or flowing, as long as there is no energy exchange with the surroundings.

Air is a fluid. Let us therefore consider a parcel of this fluid air on the sea surface. Its total energy is made up of several components. Here, we need consider only the four main ones—potential, thermal, pressure and kinetic. Others are insignificant, hence can be ignored.

Potential energy of air is due to gravity. On the sea surface, potential energy of air stays constant, whether still or moving, because gravity is constant. Therefore, we can disregard the potential energy component.

Air temperature is the measure of its thermal energy. On sea surface, at any given time of the day, we can assume that air temperature stays constant, whether the air is still or moving. Therefore, we can disregard the thermal energy component too.

That leaves us only two energy components to contend with, kinetic and pressure. Kinetic energy is a function of speed. When air is still, its kinetic energy is zero. Then there is only pressure energy to consider. Still air exerts this pressure energy perpendicular to the sea surface that it is in contact with, which in fact is the familiar atmospheric pressure.
Atmospheric pressure has also nothing to do with the weight of the atmosphere above, as it is generally understood. It has to do only with the

energy of air, which in turn depends on the energy of individual air molecules and the total number of molecules present there. The atmospheric pressure at the sea level is more than at any level higher, because there are more air molecules at the sea level, and also the molecules have much higher energies. Incidentally, the atmosphere does not get heated up directly by the sun, but by the heat reradiated from the earth, hence the air molecules at lower levels have much higher energy than those above.

When the air on the sea surface is still, that is, there is no wind blowing, the atmospheric pressure acts on the sea surface, uniformly and perpendicularly from above. Therefore, on a windless day or night, the sea surface stays flat, glassy calm, like a mirror. In other words, without the wind blowing there can be no waves at sea. Every mariner knows that. On a sunny day, the glare from the flat sea surface may be uncomfortable. But one must spend long enough time at sea to observe such wonders.

When the air starts to move, that is, when the wind gets blowing, the kinetic energy increases, steadily. Why does the wind blow or must blow are some of the questions that I shall answer towards the end of this story, in the Chapter 45: Wind Basics.

Therefore, as per the Bernoulli's law, to conserve the energy of the air parcel as it blows along as wind, somehow, and so its kinetic energy increases, the pressure energy, that is, the pressure it exerts on the sea surface must drop. In other words, as the wind blows, there is a corresponding drop in the atmospheric pressure acting on the sea surface. This drop in pressure sucks up the sea surface. The surface thus swells up into a crest, the crest of the first wave. Immediately, due to incompressibility of water and level restoring force of gravity, the waves are propagated all around. The process of wave generation by wind is as simple as that.

Fetch is the sea area over which the wind blows in a particular direction, though not quite steadily, for some duration, to generate waves. It may be several kilometres across and in length stretch from few kilometres to several thousands. Within the fetch, because the wind is blowing continuously and rather unsteadily, countless crests are constantly generated on the sea surface, at every point, and quite randomly.

The crest thus generated at each point on the sea surface radiates as waves, concentrically, due to the effects of incompressibility and gravity. The concentrically spreading waves from each point then interfere with others, either destructively or constructively. Destructive interference diffuses the waves, while the constructive interference enhances them.

The waves thus generated at each point on the sea surface and their interferences are divergent processes or processes that become complex with time. Therefore, as the wind blows, the sea surface becomes chaotic with time. In other words, waves in the fetch are erratic, irregular and unpredictable. Therefore, the waves measured at one point, say, using a wave-rider buoy, will not correspond to any other. It is therefore impossible to derive a mathematical equation to represent waves in the fetch.

The wave data collected offshore using a wave-rider buoy has to be made reasonably smooth stochastically, that is, using statistical procedures to derive the mathematical wave equations. Only then can it be used to simulate waves on a wave model. Such smooth, regular waves are hardly representative of the actual scenario offshore.

From the fetch, the waves do progress beyond into the areas of sea that are sometimes windless. There the waves acquire a regular form, as shown in the Figure 10.1 above. Such a regular wave on the windless sea is known as swell wave or swell.

At sea, as the wind-speed increases, the drop in pressure on the sea surface also increases. That in turn forces the wave-crests to be sucked up even higher. The waves therefore get taller as the wind blows faster. This relationship was observed, back in 1805, by a British Rear Admiral, Sir Francis Beaufort. He came up with a scale from 0 to 12 that connected the wind-speed and wave-height. It has come to be called the Beaufort's Scale. See the table below. It is today a standard used by mariners, meteorologists and all others in the coastal and marine fraternity.

When wind blows, the drop in pressure only acts perpendicular to the sea surface. Therefore, the sea surface can only oscillate up and down. Contrary to the general belief, the wind cannot exert any horizontal pressure on the sea surface. In other words, there is nothing called the wind pressure acting on the sea surface!

As a corollary, wind cannot make water to flow along in the direction it blows, no matter how hard it blows. That means there is no flowing water associated with wind made waves! The water only flows in a wave model, because the paddle must tread the water to make waves.

Beaufort's Scale				
Beaufort Number	Descriptive Term (Wind)	Wind Speed (knots)	Wave Height (metres)	Descriptive Term (Sea Surface)
0	Calm	<1	0	Mirror
1	Light Air	1-3	0.1	Ripples
2	Light Breeze	4-6	0.2-0.3	Small Wavelets
3	Gentle Breeze	7-10	0.4-1	Large Wavelets
4	Moderate Breeze	11-16	1-1.5	Small Waves
5	Fresh Breeze	17-21	2-2.5	Moderate Waves
6	Strong Breeze	22-27	3-4	Large Waves
7	Near Gale	28-33	4-5.5	Very Large Waves
8	Gale	34-40	5.5-7.5	Moderate High Waves
9	Strong Gale	41-47	7-10	High Waves
10	Storm	48-55	9-12.5	Very High Waves
11	Violent Storm	56-63	11.5-16	Exceptionally High Waves
12	Hurricane	> 64	>14	Phenomenal Waves

Before I move on, let me tell you something about the phenomenon of storm surge that I left unexplained earlier, which the experts believe to be caused by wind pressure heaping up water along the coast. Like the wind made waves, even the basis of storm surge is Bernoulli's Law.

As per the law, when wind blows over any surface, there is a drop in pressure on that surface. It does not matter what the surface is—sea, coast or land. In a desert, as the wind blows, the drop in pressure first sucks up the sand, which is then blown along. As the wind-speed increases, more sand gets sucked up and is blown forward. Soon it is a sandstorm. We can observe that happening even on a dusty road. As wind starts to blow, dust, along with dry leaves, bits of paper and polythene bags first float up, and are then blown forward.

Similarly, on the coast, when strong winds blow, say, when a cyclone makes landfall, the drop in pressure sucks up the sea surface along the entire width of cyclone, like a gigantic vacuum pump, with an intake several kilometres across. That pushes up the sea level along the coast to create the storm surge. In other words, the storm surge is not due to any wind pressure heaping up water along the coast, but due to the water actually being sucked up by the drop in pressure along the coast.

When the sea level thus rises above the adjoining landscape, water rushes inland, like a deluge. That causes most havoc when a cyclone makes landfall. As in the case of wave generation, faster the winds greater the drop in pressure, hence higher the storm surge. The 12 metre high storm surge during 1999 Odisha super-cyclone was due to winds blowing across the coast at more than 250 kmph!

The storm surge also must happen at sea, but is difficult to observe. Probably, an accurate GPS aboard a vessel passing through the storm could indicate the rise in sea level. But during the storm, the officer on the watch may be more concerned about the vessel's safety than observing the rise in sea level.

CHAPTER 11

WAVE FICTION

A wave, in general, at sea or anywhere else, has three basic dimensions—wave-length, wave-height and wave-period. But these dimensions may mean quite differently for a wave at sea. So we must first try and understand these dimensions in the context of a wave at sea. That may be necessary to make sense of some familiar, yet fictional aspects of the wind made waves, which the experts have come to believe as the truth.

In the Figure 10.1 above, it does appear that the wave-length is an easily measurable dimension. But there are no means to measure the wave-length of a wave at sea. We can at best infer it from a snapshot of the sea surface taken from an aircraft or a satellite. The wave-rider buoy cannot measure the wave-length. It is only meant to bob up and down, where it is moored. But in a wave model we can, because we can reach out and do so with a measuring rod.

The wave-height is generally accepted as the vertical distance between the highest point of a wave, the crest, and its lowest point, the trough. That is what has been indicated in the Figure 10.1 above. But for a wave at sea, that may not be quite right either. A wave at sea is the result of vertical motion of a series of points on the sea surface. Therefore, the wave-height is the maximum displacement of each point on the sea surface as it goes through one oscillation. The wave-rider buoy only measures this displacement at the point, where it is moored.

For a mariner at sea, only the wave-height matters, because it affects the comfort onboard and also safety. He therefore observes the wave-height at regular intervals, usually once every four hours. For the observation he has at his disposal no instrument. He estimates the wave-height by comparing the rise and fall of sea surface against some feature of his ship. That is highly approximate. But that is the wave-data that reaches the meteorological offices, world over, often months later. That still makes the bulk of wave-data currently available worldwide!

The wave-period, in a general sense, is the time taken for a wave to travel through one wave-length. But for a wave at sea, that cannot hold true, because it is only a vertical oscillation, without any horizontal component of motion, like most other waves in nature, both visible and invisible. Hence there can be no relationship between the wave-length and wave-period. Wave-period therefore should be the time taken by a point on sea surface to oscillate 'once' from highest to its lowest position or the other way, that is, 'once' through the wave-height. But as a convention, it is accepted as the time taken to oscillate 'twice' through the wave-height, because for all other waves, wave-period is the time taken for a wave to travel through one wave-length, that is, the wave traversing from one crest, through a trough, to the next crest, during which it goes 'twice' through the wave-height.

Only flowing water can move sediments, in suspension or along the bed, provided it is swift enough. Because the experts believe that the waves move sediments, they also must believe that the water in the wave flows along or the wave itself moves forward, like an advancing column of oscillating water. If the wave or the water in a wave is moving forward, it must do so at some speed. They call it the wave-speed or wave-celerity or speed of the wave form.

It is impossible to measure the wave-speed at sea. But it can be observed in a wave model, because the paddle, besides making the waves, also pushes the water column forward at some speed. It may also then be possible to measure that. Nevertheless, the wave-speed is normally computed for solving the wave equations. The formula used is so simple that it does seem quite right.

$$Wave\text{-}speed = Wave\text{-}length \div Wave\text{-}period.$$

Speed of a body is rate at which it changes position. In other words, the distance it travels in unit time. Water does not move horizontally with the wave at sea, but only vertically. So, there is nothing called wave-speed for a wave at sea. The waves only appear to move. It is only an optical illusion! That makes the above formula a mathematical fiction.

But it goes to support the theory that the waves move sediments. Wave-length of a wave at sea, though not amenable to real-time measurement, is usually of the order of hundreds of metres and wave-period in the order of few seconds. Therefore, the wave-speed computed by the above formula yields a fabulous value. At such speed, moving the sediments in suspension or otherwise may not be a problem. If such speeds are somehow possible at sea, there may be little need for propulsion to sail the seas. We can always hitch ride on a wave, as Elise in the fairytale was hoping to do. Nature does not offer free rides, even to the sediments!

Besides the wave or the water in the wave moving horizontally on the sea surface at some wave-speed, the experts in wave dynamics also claim that the water traces a circular path along the vertical plane. They call it the orbital motion of waves. Obviously, they have only observed the motion in a wave flume.

Wave model cannot be used to study wave dynamics. That is done on a wave flume. It is an oblong tank, usually with sides of glass, in which the waves are simulated using a motor-driven paddle, just like in the wave model. Besides being used to study wave dynamics, the flume is used to test the models of prototype ships for manoeuvrability and sea-keeping qualities and also to test the models of coastal and marine structures that would be exposed to waves, for strength and durability—seawalls, breakwaters, piers, jetties and others.

Actually, the orbital motion is only possible in the wave-flume, and nowhere else, certainly not at sea. The paddle in a wave flume turns around its axis. Therefore, besides making waves and pushing the water column forward, it also imparts a 'spin' to the wavy water column. In other words, it is the paddle that sets off the orbital motion. It is impossible at sea, because wind does not tread the water to make waves.

Also from observations in the wave flume, they have established that the wave's orbital diameter equals its wave-height. So the formula for the speed of orbital motion or orbital-speed is as follows.

$$Orbital\text{-}speed = \pi \times Wave\text{-}height \div Wave\text{-}period.$$

This formula too is a mathematical fiction, because the orbital motion is impossible at sea.

That brings us to another common parameter of a wave at sea that is entirely fictional, the wave direction. Wind blows in a particular direction for a while, at least, before changing to another. Even though the wind makes waves, it is impossible to tell the direction of waves. The waves at sea are no different from the ripples on a pond. When we create the ripples by tossing a pebble, they radiate in every direction, in concentric circles. The 'apparent' direction of the ripples depends on where we are observing it from.

On the coast, the waves are either head-on or at an angle. That depends on the lay of coast with respect to the fetch. Therefore, wind direction could only serve as an indicator of the 'apparent' direction of waves. But for that, there is no way of telling where the waves are heading to or coming from.

Actually, the question of direction does not arise when there is no horizontal movement. It is impossible to tell the direction of something that does not move horizontally. The waves do not go anywhere, only up and down. But the illusion of motion is so strong that everyone believes the waves are heading to or coming from a particular direction.

The sediments are denser than the seawater, between 2½ to 4½ times denser, depending on the mineral composition of the rocks from which they originated. The sediments must therefore remain on the seabed, held down by gravity. How can the waves on the sea surface move the sediments that must stay on the seabed?

If you have doubts about the waves being surface phenomena, you must talk to the submariners. When the sea gets rough, they would happily dive down to enjoy smooth sailing.

The question does not deter the experts. They feel that the waves are capable of churning the dense sediments on the seabed, bringing them to the surface and then moving them along in suspension. So they claim that the wave's orbital motion continues below the surface, but with the orbital diameter diminishing with depth. The depth at which the diameter reduces to zero is known as the wave-base, therefore dependent on the wave-height. Taller the waves deeper is the wave-base!

If the seabed is shallower than the wave-base, the experts believe that the waves on the surface can actually churn the sediments on the bed up to the surface. Hence, the wave-base is sometimes defined as the greatest depth at which the sediments on seabed can just be stirred by the wave above[5].

In a wave flume, it may be possible to observe the orbiting wave-water interacting with the bed. But even there, it is impossible to simulate the stirring of dense sediments on the bed, let alone bringing them to the surface.

There is no way the wave-base theory can be validated at sea. Water samples collected near the coast in the wavy waters will not yield sediments, anywhere. No sea swimmer has ever encountered sediments churned up by the waves, when swimming near the coast through wavy waters. You can check that yourself. Do not go by the colour of water. Water near the coast may appear greenish brown, particularly along the Indian Coast. That is not

---

[5] c.f. Entry 5928, Hydrographic Dictionary, IHO Special Publication No. 32 (Fifth Edition)

because of the sediments held in suspension, but due to the chlorophyll content of some microscopic marine organisms in the nearshore waters. So, the theory of wave-base too is a fiction.

With that we can now take on the most popular of all coastal fiction—waves move sediments. Can it really do so? From the discussions so far, the question may seem unnecessary. But the experts believe it rather dogmatically. In fact, great deal of scientific effort has also gone into it. Therefore, we must dwell on it till we can find conclusive argument either for or against.

One of the factors involved in wave making is gravity. If there is no gravity, water would not keep level. It would take any shape we give it, which would remain until altered by some other force. That means without gravity, the sea surface cannot oscillate as waves.

In addition to making waves, gravity holds the dense sediments down to the seabed. Without gravity, the sediments would float up with slightest of provocation, and remain floating, not only in water, but also in the air above.

Therefore, with the ubiquitous force of gravity, if the waves are picking up sediments from the seabed, due to orbital motion or due to any other esoteric phenomenon, placing them in suspension and then moving them along, logically, it is gravity working against itself. Gravity makes waves. Gravity holds the sediments down. Therefore, gravity is picking up sediments from the seabed that it is naturally bound to hold down. That cannot happen on the Planet Earth! It is therefore impossible for the wind generated waves at sea to move sediments, anywhere.

But all that changes when the wave meets the coast. There it breaks. We must therefore observe what happens to a wave when it breaks and how it behaves thereafter. Beach is an ideal place for the observation, because the waves break there almost always. Waves also break on a rocky coast, but there it can be dangerous. Even on a beach, the observation is possible only on a calm day. When the sea gets rough, it may get dangerous even on a beach.

CHAPTER 12

BREAKING WAVES

A popular theory in wave dynamics suggests that the waves break, where the depth is less than or equal to 1.3 times the wave-height. This seemingly precise relationship is said to have been based on wave studies done during the World War II by the military commanders engaged in amphibious operations.

How they may have arrived at this seemingly precise relationship is not difficult to establish. We can be certain that it could not have been based on observations made on the coast, because measuring the depth, where waves break is impracticable, even with modern day instrumentation.

If the wave is only a vertical oscillation on the sea surface, with no horizontal component of motion, there is no a reason why the relationship should hold true anyway.

A non-flowing, vertically oscillating wave can sustain itself only if there is sufficient depth for a complete oscillation. That is nothing quite esoteric. That is the depth where the wave's lowest part, its trough, touches the bed, in other words, where the depth is less than or equal to half the prevalent wave-height.

On the other hand, if the water flowed along with the wave, as it does in a wave flume, then in shallow depths, slightly before the wave-trough touches the bottom, the flowing water would engage the bed and slow down, due to what is known as the 'bed shear', a sort of friction experienced by water, when it flows close to the bed. The water on the surface of the flume however continues to move forward without much slowing down. As a result, the wave on the surface would lunge forward and appear to collapse before its trough touches the bed.

There is no water flowing associated with the wind generated waves, therefore the question of slowing down due to the bed shear does not arise. We can therefore infer that the military commanders studied the wave breaking in a wave flume and nowhere else.

Where on the beach the wave breaks would depend on two factors: wave-height and depth. Depth along the coast varies with tide. For a given wave-height, the waves would break seaward at low tide than it would at high tide. As the tide rises, the wave breaking line would steadily shift landward, provided there is no change in the wave-height. As the wave gets taller, it should break more seaward. But tall waves are usually associated with strong winds, which in turn cause the storm surge. As a result, the depth along the coast increases. And so the waves though tall would break landward.

What is important to us is not where the wave breaks, but what happens after it breaks. We can observe that on a beach. Before we do so, we must note the angle the wave is making with the coast. If the wave is at an angle, it would begin to break at the end it meets the beach first. Then rest of the wave would tumble after, progressively, like a domino. That is a common sight on the East Coast. When the wave is head on, it breaks in one go along the length of the beach, provided it is a straight one. That usually is the case on the West Coast during the monsoon.

As the wave breaks, water held up in its crest, which let us call the wave-water, tumbles down into a frothy and turbulent mass of water that goes surging upslope along the bed. But the froth soon clears up. Thereafter, it is only a sheet of clear water flowing upslope. Because it is flowing upslope, it slows down rapidly and stops. As it stops, it may appear to seep into the sand. Let us call the wave-water flowing upslope the run-up. What is important is not how far up the beach the run-up goes, but only its direction. The run-up invariably would flow along the same angle the wave made with the beach before breaking. If the wave was at an angle, it would flow both upslope and along the beach. As the angle gets broader, it would flow more along the beach, than upslope. On the other hand, if the wave was head on, it would only rush straight upslope.

After the run-up stops and seeps into sand, almost immediately, the wave-water flows back to sea, literally, seeping out of the sand. Let us call this wave-water flowing back to the sea the run-back. Like the run-up, even the run-back flows along the bed, like a sheet of water, but with little froth. The run-up is powered by the momentum of the collapsing wave-crest, like a bucket of water flung on the floor, whereas the run-back flows down the beach gradient, powered by gravity alone.

The run-back therefore starts slowly and picks up speed as it heads downslope to sea. Here too, the important aspect is its direction. No matter the angle the wave made with the beach, the run-back will only flow perpendicular to it, that is, straight down the beach. At the end of its run downslope along the beach bed, the run-back merges with the sea, and ceases to flow. Say, if it did not stop flowing on reaching the sea, it would have given rise to a current of water flowing seaward. While that may seem quite plausible, there is no empirical evidence of such a current. We shall soon see that nature does not require such a current to drive its processes further on. It is better equipped with the ubiquitous force of gravity.

To sum up, after the wave breaks, it is flowing water, first the run-up, then the run-back. Unlike the vertically oscillating water in the wave at sea, the flowing water on the coast can do a lot of work, even move sediments.

CHAPTER 13

COASTAL SEDIMENT DYNAMICS

Nearly everyone is under the impression that the waves and tides, sometimes even the winds power the coastal sediment dynamics, the natural movement of sediments on the coast. Actually, there are six agents involved, either directly or indirectly—gravity, wind, hydrostatic pressure, runoff, run-up and run-back. The last two are however from the waves. Therefore, we can say that the waves too contribute to coastal sediment dynamics, but indirectly.

Gravity works on the coastal sediments, like it would anywhere on the land, moving them down from the higher level to the lower. The sediments do roll down a steep coast aided by gravity alone, perhaps with a slight nudge from humans or animals or even wind.

Wind may blow a bit of sand on the coast, now and then, but its real contribution towards coastal sediment dynamics is quite indirect. I shall deal with that presently, when I discuss the formation of berms and dunes on the coast.

The hydrostatic pressure acts on the beds of rivers and lagoons near the coast, to push the sediments from their beds through the intervening soil to the coast. See the Chapter 21: River through the lagoon and the Chapter 26: Road to Chellanam. In some exceptional cases, the sediments can even be pushed out so forcefully that they could erupt from the seabed offshore. You will encounter an interesting case later in the story, in the Chapter 44: Mystery of the Dead Coast.

Runoff or flowing rainwater also moves sediments. That is a common sight on land. Sediment-laden overland runoff reaches the coast mainly through the natural channels of streams, rivers or creeks. In coastal regions that receive heavy rains, like the West Coast during monsoon, the runoff sometimes flows directly across the coast, which on a beach normally seeps unseen through the sand, and on a muddy coast, it flows on the surface, but again quite indiscernibly.

The main agents of sediment dynamics on a beach are run-ups and run-backs. Unceasing run-ups and run-backs make the beach the most dynamic place on the earth's surface, particularly, when it comes to the movement of sediments. Let us therefore examine the action of run-ups and run-backs on the beach sediments. Though the run-up occurs first, let us first study the role of run-back, because it is simpler to do so.

The run-back, we have already noted, flows straight down the beach to sea aided only by gravity. With that in mind, let us begin a round of observations, on a beach with a gentle gradient, and on a calm day.

The general view is that the sediments are moved only when the conditions on the beach are rough, in other words, when the big waves come crashing on the beach. Let us see if the same happens when the conditions are calm, which is the situation most of the time on any beach.

Walk up barefoot to the waterfront. Spend few moments standing still in about ankle-deep water. Let the small waves lap your feet, one after another. Stand as still as you can. Sense the feeling. You are sinking! Sand seems to be slipping away from under your feet. It may make your soles feel ticklish. What is making you sink? Or, why is the sand slipping away?

You can be sure it is not because of your weight alone. You are sinking because the sediments under your feet are being whisked away to sea. What may be driving the process?

Take a few steps back, upslope. Choose a spot awash only by the froth-free run-up, hence fairly upslope. After the run-up passes upslope, before the run-back flows over, when the bed is briefly exposed, scoop out some sand, with a beach shovel, to make a small pit. Soon you see the run-back flowing over the pit, filling it up with water. Observe closely what is happening to the pit. As the run-backs flow over it, one after another, the sediments from upslope flow into it. Few run-backs later, there is no trace of the pit. That means the run-backs are steadily moving sediments from the beach to sea. Not convinced?

Here is another proof, a clinching one yet. Observe the bed, close to the waterfront, after the run-back passes into sea. You see 'clear' water flowing through innumerable tiny gullies or valleys, like little meandering streams flowing into sea. You see no sediments flowing in these streams, but only clear water. But these streams are about the best evidence of sediment removal from the beach with every run-back. The evidence is in the tiny valleys through which the clear water is flowing to sea. The valleys were created only a while ago when sediments were removed to sea by the run-back that just passed to sea. The run-back on its last leg is quite swift, being gravity driven.

The valleys do not remain for long. Even as you watch, you see them being filled up with sediments slipping down from the upslope. Only to be created afresh as the next run-back flows over. The valleys are therefore constantly created and obliterated, in a random manner.

We can therefore deduce that on a beach, the run-backs steadily and unceasingly remove sediments from the beach to sea. Sediments can be moved only by flowing water. Run-back is water flowing under gravity, like it is anywhere on land. It flows not due to any motion imparted by the waves. Wave breaking only delivers water to the beach. That happens whether the waves are small or big. Obviously, when waves get bigger, the amount of water delivered to the beach is also more. That in turn translates as more water flowing back to sea with each run-back, hence more sediment is moved to sea each time.

The run-back being gravity driven and because of its brief run does not gain much momentum before merging with sea. It is therefore capable of moving only fine sediments, clay and silt, occasionally some fine sand.

Therefore, with every run-back clay and silt are removed to sea from the beach. That goes on unceasingly. As a result, the coarser sand remains on the beach. That is why beaches are sandy, everywhere!

On a sheltered coast, there are no waves breaking. Therefore, there are no run-backs to remove the clay and silt. A sheltered coast therefore stays muddy, always. Take a look inside the Mumbai Harbour. There is no sand anywhere, only mudflats. That is the proof that the harbour is sheltered from the waves.

On the north coast of Gulf of Khachch, the nature of sediments becomes progressively becomes fine, as we head from west to east into the gulf. At the gulf's western end, where reasonably big waves break on the coast, it is sandy. The waves get smaller as we head into gulf. As a result, the sediments become finer. Deep inside the gulf, it is entirely sheltered, hence muddy.

Recall the case of the sand-trap used to measure longshore drift that we discussed earlier, in the Chapter 6: Mythical Currents. What fills the trap is not the sand that the mythical longshore currents move along the coast, but clay and silt that the run-backs bring directly down from the coast across the trap.

Let us now study the action of run-up on the coastal sediments. The run-up is what follows the wave breaking. Immediately after the wave breaks, the run-up is frothy and turbulent. But that soon clears up. Then only 'clear' sediment-free water can be seen flowing upslope, which rapidly slows down, stops and seeps into sand. Therefore, if the run-up is doing any work on sediments, it must be immediately after the wave breaks, in that frothy turbulent phase. But whatever happens in there is almost impossible to observe. Moreover, it is over too soon. There is little time to carry out any systematic observations, with or without instruments. Actually, there are no instruments designed for such an observation. On second thought, we may have just the right kind of instrument, perhaps the most sophisticated and sensitive one there is—our body!

Our aim is to observe whatever may be happening within that brief frothy turbulent phase that immediately follows the wave breaking. For that, only our body is equipped with the necessary sensory devices, provided we put ourselves there and pay attention.

So let us go to a beach, for yet another field observation, using only our body as the instrument of observation. Choose a day when the waves are modest, not too rough, when it is unsafe to venture on the beach, and not too calm, when they are but ripples.

For this observation, you must be gutsy. At least, you must not be afraid of being swamped by water. For safety, you must have a buddy standing by you. You must have a nose-clip on, one that a scuba diver wears to prevent breathing in through the nose. You must also don a pair of underwater goggle. Nose-clip and goggle are necessary to prevent sand getting into your nostrils and eyes. Also, remember to keep your mouth shut when you are doing the observation, lest you end up with mouth full of sand. This observation may seem more like a stunt. Actually, lot many stunts do go on in the name science anyway, why not one more?

Before you begin, note the angle the waves are making with the beach, and where they are breaking. Also, make sure there are no obstructions, rocks or other debris around, which may injure you, particularly along the run-up. Remind your buddy to keep a close watch over you and to bail you out, if he or she finds you in some confusion or difficulty. Sometimes you could end up getting choked.

Get close to where the waves are breaking. Face an oncoming wave. Just as it is about to break, take a deep breath, and drop backward, quickly and lie down on your back, with your feet into the breaking wave. You have to be very quick. The wave soon crashes over you. You are now within the frothy turbulent phase of the run-up. It is quite a sensation, perhaps scary too, as the sand-laden water surges past you. You feel the sand scraping past you. Some sand may even gather up between your legs, armpits, and even around the neck. It may be quite uncomfortable. But then, there is no other way to experience the rush of sand with every run-up.

You may also find yourself floated up and pushed upslope, along with the sand. After all, your body is not as dense. Your buddy can tell the direction you were pushed. It will be same as the run-up.

The observation was over too quickly. So the process may be still unclear. Let us therefore perform another experiment to make things clearer. Do not worry. It is not a dangerous stunt. It is a thought experiment.

Albert Einstein was famous for his thought experiments. He probably performed countless when studying the motion at speeds close to that of light. He could not have done that in any laboratory, so he did it in his head. He probably took a ride on a photon, the particle of light, to see for himself, whatever may be happening at such speeds. It may be difficult to say what exactly he saw or experienced, because anything is possible in imagination. There are no physical limitations to contend with. It was after such thought experiments did he come up with the landmark Theory of Relativity.

The theory incidentally could not have been evolved from the then existing body of knowledge in physics by applying any kind of known logic or through any empirical evidence thereafter until today. It is a product of pure imagination of highest order or of a thought experiment of extraordinary brilliance. That is what makes Einstein such a great scientist. But for him, the theory of relativity may have still eluded the physicists.

In our thought experiment, we shall however be hitching a ride on a sand particle. Close your eyes, though it is not necessary, but helps keep out distractions. Imagine yourself clinging to an average sized sand particle, normally found on a beach that you may have visited. When you have got a good grip of the particle, imagine a wave breaking right over you. You have experienced it, therefore imagining it may not be difficult. There is a sudden surge of water all around you. Almost immediately, you find yourself tossed up and pushed forward by the gushing water, a motion more topsy-turvy than any roller coaster in the world.

But the tossing, tumbling motion soon comes to an end, almost as soon as it started. Thereafter, you find yourself being rolled along the bed along with the water that is still surging upslope. As you are rolled along, you see other sand particles also being rolled along, either side. Even this ride along the bed ends soon. You come to rest, but the water continues to flow on. It has slowed considerably, because it is flowing upslope. It no longer can move you any further upslope. Other sand particles that rolled along with you too stop, like the marching troops called to halt.

Though you do not get to see it from where you are stopped,

clinging on to your sand particle, the water that flowed upslope, leaving you behind, travels some more, halts and seeps into the sand. You know the process. In a while, the water flows back to sea, downslope and perpendicular to the coast, as run-back. Now there are two cases to consider, one, when the wave meets the coast at an angle, and two, when it meets head on.

When the wave is at an angle, your initial ride is also along the same angle to the coast. It therefore takes you both along the coast and upslope. Broader the angle of run-up, more you will travel along the coast. The ensuing run-back flows back to sea at some distance ahead of where you halted. You do not encounter it. But, as you wait, a run-back does flow right over you. It is of a wave that broke slightly upstream. This run-back has no strength to move you, because you are on a relatively heavy sand particle. Even the other sand particles that halted with you stay put, unmoved. But in this run-back, you see some clay and silt being moved along. It will carry more as it picks up speed on its journey to sea.

In the case two, when the wave is head on, you find yourself only moved straight upslope. You cover less ground than the former. Run-up loses steam rapidly, because it is flowing straight up against the gradient. Soon, it leaves you behind and flows upslope some more. There it seeps into sand, only to flow back to sea, right over you. As in the previous case, even this run-back is not strong enough to dislodge you. Here again, you see small quantities of clay or silt being carried past you.

In both cases, after you are stopped, you find little action around you except weak run-ups and run-backs flowing past. But as you wait, you are joined by more sand particles, with every wave that breaks downslope. As you wait, clinging to your sand particle, the tide rises. Soon another wave breaks right over you. You are again hurtled forward, as before, upslope and along or only upslope, depending on the angle wave made with the coast. In time, the other sand particles join you.

You are therefore pushed upslope in stages, till the tide rises no more. That happens at spring tide, full or new moon. Your movement upslope therefore may take up to a fortnight. When you have reached the top of your journey, riding the sand particle, you may now get off and return to the real world. The particle that you rode will remain at the beach head, around the high water line, until trampled by a human or animal or moved by wind and gravity and returned to the wave-breaking zone. Then again it will be pushed up, with the very next wave breaking over it.

Therefore, when the waves break at an angle to the coast, the sand is pushed up and also along the coast. Sand therefore moves right on the coast and not through water in suspension or otherwise borne by any current flowing along the coast. The drift of sand, which we have come to call longshore drift, is powered by the run-ups. It is a motion in small steps, and not as a flowing current. Broader the wave angle, greater the movement of sand along the coast.

On the other hand, when waves are head on, sand is only pushed up the coast, once again in small steps, and none whatsoever along the coast. Therefore, the longshore drift is possible only when the waves meet the coast at an angle.

Normally, the sand reaches only up to the high water line, which is slightly higher than the spring tide level, because of the run-ups surging past that level. Therefore, the beach should have extended only up to the high water line or marginally beyond. But almost everywhere, where the terrain is reasonably flat, we find lot more sand landward of the high water line, often deep inland and at a much higher level. The sand is sometimes heaped up as a berm or dune. For a recap, berm is a ridge of sand along the beach, whereas the dune is a hillock of sand. Both are usually landward of the normal high water line, standing well above the normal high water level.

Many believe that the berms and dunes are the handiworks of wind alone, as in the case of a desert, where they are quite common. Strong winds do blow across the coast, but by themselves do not heap up the usually wet sand to form berm or dune. Nevertheless, the winds do play an important role, though not directly as in a desert.

When strong winds blow across the coast, seasonally or otherwise, sea level at the coast goes up due to the storm surge. We have discussed that earlier. During the storm surge, the depth along the coast also increases. As a result, the big waves break further inshore. The resulting run-up drives the sand beyond the normal high water line, to deposit as berms or dunes.

When waves are at an angle to the coast, sand gets deposited as a dune. Dunes are therefore common on coasts frequented by storms that make landfall at an angle, for example, the East Coast, where the cyclones that start south of Bay of Bengal make landfall at an angle to the coast.

On the West Coast, during the monsoon, strong winds blow head on, so do the waves meet the coast head on. The onshore winds give rise to a significant rise in the sea level all along the coast. The head-on waves therefore break further up the coast to build up the berm.

After a while, both berms and dunes are razed down and the sand is spread around. That is how there is so much sand landward of the high water line.

Run-ups only move the sediments that are already on the beach. Run-backs however steadily remove the sediments from the beach, usually only fine ones. That, in time, would render the gradient of the coast so steep that even coarser ones would roll down to sea, with little provocation, aided only by gravity. In other words, the beach begins to lose sediments. That is what erosion is all about, beach losing sediments. But not every beach is naturally eroding. That means it is being replenished. The obvious source is land, being at a higher elevation.

Land is being eroded, unceasingly. Several agents are at work—solar heat, falling rain, flowing water, freezing water, sliding ice, reacting chemicals, scarping dusty winds, relentless roots, burrowing animals, and the biggest of all, humans. Humans cause land erosion in more number of ways today than ever in the past—deforesting, farming, irrigation, mining, urbanising, landscaping, building road and rail networks, laying underground pipelines, and many more. The list is potentially endless.

Land erosion creates sediments—boulders, cobbles, pebbles, gravel, sand, silt and clay. Flowing water and sliding glaciers move these sediments overland. These are powered by gravity. Gravity on its own also moves sediments, through landslides and mudflows. Winds too move sediments. The sediments eventually reach the sea, sooner or later. Even during the movement seaward, coarser ones are ground fine by mutual abrasion. Heavier ones—boulders, cobbles, pebbles and gravel—remain inland until broken down further. Only sand, silt and clay normally make it to the coast, thence to sea.

Runoff is the principal mover of sediments to the coast, flowing either as sheets of water or through natural channels of streams, rivers or creeks. Sediments reach the coast, as a general rule, through these channels.

It may seem that the channels are delivering sediments directly to sea. If that was so, the coast either side of the channel mouths would get no sediment supply, therefore erode. But that is not happening. That means the sediments discharged by the channels are being deflected coastward. Let us see how the nature does that.

CHAPTER 14

SANDBARS AND SPITS

Sandbar is an underwater ridge of sediments, though not entirely sandy, that forms across the mouth of a river. Sandbars are common where the waves are generally head on to the river mouths, as on the West Coast, during the monsoon. That is also when the rivers discharge sediments copiously.

For the sandbar to form, sediments discharged by the river must be deposited slightly offshore of its mouth, in a narrow band. That is possible only if the sediments, which are invariably flowing along the bed, slow down abruptly. If not, the sediments would spread on the seabed over a large area, like a fan.

The monsoon waves engage the river's discharge head on, but only on the surface. So, only the surface layer of discharge slows down, whereas the sediments are flowing along bed. The river's discharge is however a homogenous mass of flowing water from the surface down to the bed. Therefore, the slowing down on surface passes down to the bed. It may take a little while though. By that time, the sediments flowing along the bed would move a little ahead offshore, and deposit in a narrow band, almost entirely spanning the river mouth. As more and more sediments deposit there, it grows into an underwater ridge, the sandbar. It will not rise above the level of sediments flowing out of the river. Hence the sandbar invariably stays underwater.

After the sandbar forms, the waves begin to break on it, provided the depth over it is less than or equal to half the prevalent wave-height. Wave breaking does not however tear it down, because of the continuous sediment supply from the river. The sandbar would therefore remain in place till the end of monsoon. When the monsoon rains cease, the river also ceases to discharge sediments. So the sediment supply to the sandbar is cut off. It then does not take long for the breaking waves and gravity to pull it down, though not entirely. A remnant usually remains in place, only to be restored when the next monsoon sets in.

Nature does nothing without purpose. What purpose does a sandbar serve? That may not be quite apparent until it is prevented. In the past, the sandbar did not bother the fishermen. They operated mostly in small canoes that could easily ride over it. Nowadays, canoes have given way to large motorised fishing boats that are either anchored within the estuary or berthed alongside the wharfs built on its banks. The sandbar made it difficult for these boats to operate, particularly during monsoon. Fishermen therefore wanted the river mouth deepened, in other words, rid of the sandbar.

The experts, probably, after hydraulic model studies, proposed building groynes both sides of the river mouth. See the Figure 14.1 below. With the groynes in place, the sandbar did not form. And so the mouth stayed deep. But what prevented the sandbar?

The groynes, in effect, narrowed the mouth. That increased the river's rate of discharge significantly. That again is Bernoulli's law in action. Head-on waves therefore did not slow down the sediments sufficiently to form the sandbar. Instead the sediments spread over a wide area, like a fan, beyond the groynes. But that came at a price—erosion both sides of the river mouth!

When a sandbar forms, the river no longer can discharge its sediment load to sea, directly. Instead, it is deflected into two streams that flow towards the coast, either side of the river mouth. The run-ups then deliver these sediments on to the coast. That is how the coast either side of the river mouth gets the sediment supply and stays stable.

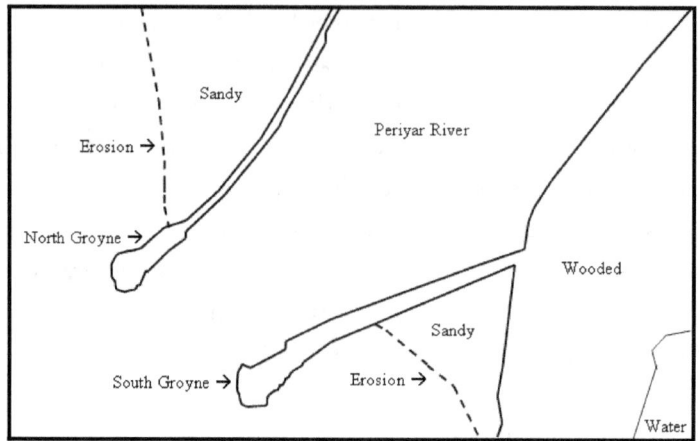

Figure 14.1: Groynes on the Mouth of Periyar River

During monsoon, in spite of heavy rains on the coast, coast either side of a river mouth does not get sediment supply directly through the runoff. In the vicinity of the river, the runoff invariably flows into it and not across the coast. The sandbar therefore is the nature's way of ensuring stability of the coast either side of the river mouth. Therefore, when the sandbar is prevented, there is no mechanism to deliver sediments there, so erosion sets in, both sides of the mouth.

Now let us examine the situation on a river mouth, where the waves are at angle. When waves meet the coast at an angle, the sediments are moved along the coast. That is the longshore drift. To sustain the drift, coast must get sediment supply. Sediment source is invariably a river mouth upstream.

To sustain the longshore drift, the river mouth must perform two functions. One, it must deflect the sediments it discharges towards the coast downstream. Two, it must bypass the longshore drift from the coast upstream across it to the coast downstream. Both are done by the spit—a visible ridge of sand at an angle to one end of the river mouth. That in effect turns the mouth towards the coast downstream.

Spit therefore sustains the longshore drift. That is its only purpose. Therefore, if the spit is prevented, the coast downstream will not get the sediment supply. So the erosion will set in downstream, almost immediately. Examples abound on the East Coast.

Like the sandbar, even for the spit to form certain conditions must be met. To start with, the river must discharge sediments, only then can the spit begin to form. Unlike the sandbar, the sediment supply from the river need not be copious or sustained. For the sandbar, river is the only sediment source, whereas spit also gets sediment supply from the upstream via the longshore drift. The longshore drift is therefore a necessary condition for the spit to form. Just as the spit ensures sediment supply to sustain longshore drift downstream, the drift from the upstream is necessary for the spit, both to form and to keep it in place. In other words, if the longshore drift is somehow disrupted upstream of the mouth, the spit will not form.

To drive the longshore drift, the waves must meet the coast at an angle. Even for a spit to form, waves must meet the river mouth at an angle. It will form at the end that meets waves first. As a corollary, the spit will not form if the waves are prevented from reaching the river mouth, say, by a breakwater or groyne.

Also, for the spit to form, mouth must necessarily be shallow. That is quite naturally so. If the mouth is dredged, the spit will not form, because the sediments discharged by the river and what comes from the upstream will slip into the deepened portion. If the mouth is not dredged thereafter, it will soon silt back to the original level. The spit will return.
Let us take the case of a small river on the East Coast. As Southwest Monsoon sets in, the waves begin to meet its mouth at an angle from south. When the river's sediment-laden discharge encounters the waves, just as in the case of a sandbar, slow down happens at the surface layer, which soon passes to the bed. Sediments flowing along the bed therefore slow down and deposit to form a 'sandbar' that extends north at an angle from the southern tip of the river mouth.

As the sandbar forms, the longshore drift from south delivers sediments on it. That makes it rise above the water level to become a 'visible' spit. If the longshore drift is arrested upstream of the mouth, the sandbar will form because of the sediment supply from the river, but continue to remain underwater without growing into a spit.

After a spit forms, it will remain in place as long as the longshore drift continues to deliver sediments, even if the river stops doing so. But if the river stops discharging enough water, for one reason or the other, longshore drift may cause the spit to grow into a sand-ridge, to block the mouth completely. Later on, whenever the discharge of water resumes, say, after a heavy spell of rain in the catchment, a portion of sand-ridge towards the coast downstream will be knocked off, making it a spit again.

Southern spit stays in place till the end of Southwest Monsoon. Thereafter, the angle of waves slowly shifts northward. During the transition, the longshore drift from south ceases, and waves, for some time, becomes head on to the mouth. That is when the south spit is torn down. By the time Northeast Monsoon sets in, the waves are from the north. That then drives the longshore drift southward. The entire process gets repeated to form a new spit at the northern tip. The process goes on year after year, until someone tries to prevent the spit.

Let us now examine the consequences of preventing spit on the mouth of river on the East Coast, the Coovam River.

CHAPTER 15

STORY OF THE MARINA BEACH

Long ago, Coovam was just a small seasonal river that drained into the Bay of Bengal. As usual, the spits formed at its mouth, north and south, alternately, depending on the monsoon. Madras Town was built on its banks. The town soon grew into city. It is now the City of Chennai, a metropolis. As the city grew, sewage and wastewater were let into the river, liberally, largely untreated. Soon, from a seasonal river, Coovam became a perennial river of stinky sewage. Spits however continued to form blocking its mouth as usual. As a result, the city's sewage did not flow out to sea, but lay trapped within. That raised a stink. The municipal authorities therefore decided to keep Coovam's mouth open throughout the year. That meant preventing the spits.

The experts proposed a groyne, to be built at the southern end, extending into sea from the river's southern embankment. The groyne was inclined north. Let us call it South Groyne. See the Figure 15.1 below. But what prompted the experts to go for only one groyne, when there was also longshore drift southward during the Northeast Monsoon?

They probably concluded that the Chennai Port, north of the

Coovam's mouth, arrested the 'longshore current' that flowed south during the Northeast Monsoon, therefore only the northerly 'longshore current' delivered sediments into the river mouth to form the spits, both north and south. There seems to be no other reason.

After the South Groyne was built, during the ensuing Southwest Monsoon, no spit formed at the southern tip, but a sandbar did. No one expected that. Obviously, they had not taken into account the sediments discharged by the river.

At the same time, the narrow beach south of Coovam began to accrete, because the groyne arrested the northerly longshore drift. Before long, the beach extended up to the groyne's seaward end, tapering off south, like a long triangle of sand. That is Chennai's Marina Beach, a bonus for trying to prevent the spit at Coovam!

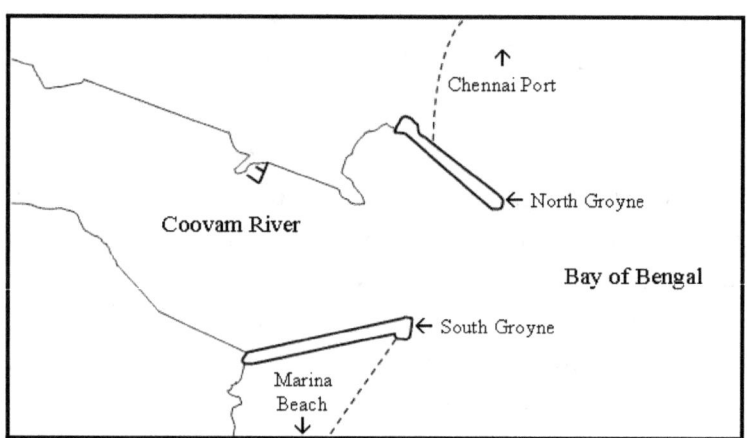

Figure 15.1: Groynes on the Mouth of Coovam River

As the Marina Beach grew seaward, the coast north of Coovam began to erode. But the erosion did not spread north, because the coast there was well fortified with a revetment or fortified seawall. That is the shoreline of Chennai Port, which once was the Marina Beach that accreted after the 1876 Madras Pier was built, followed by the Madras Harbour. See the Chapter 6: Mythical Currents.

The Southwest Monsoon passed without the spit, but only the

sandbar. Often, the sandbar spanned the entire mouth, like an underwater sand-ridge. Normally, when the waves shift from south to north, with the change of monsoon, for a while, they are head on to the river mouth. That is when the southern spit gets torn down. The head on waves should have torn down the sandbar too, but did not, because the north-inclined South Groyne shielded most of the river mouth from the head on waves. Therefore, most of the sandbar stayed in place even when the Northeast Monsoon set in.

The waves from north, contrary to expectations, did drive the longshore drift southward. That also is a proof that there is nothing called 'longshore currents'. What drives the longshore drift is the run-up of waves breaking at an angle to the coast. The sediments thus driven south began to deposit on the sandbar, which was already in place. As a result, the north spit formed rather quickly. Often, the spit grew into a sand-ridge blocking the river mouth completely! Even when the next Southwest Monsoon set in, a portion of this north spit remained.

The experts were again called in. This time they suggested building another groyne, the North Groyne. See the Figure 15.1 above. But even with the North Groyne, north spit continued to form during Northeast Monsoon. That means the groyne did not arrest the longshore drift from north. The reason is simple. The North Groyne, unlike the one south, was not an extension of the river's embankment, but stood out from the beach, like an island. Why they chose to do that defies logical explanation. During Northeast Monsoon, the sediment-laden run-up surged past the groyne's root, its landward end, during the high tide. Thus the run-up driven southerly longshore drift continued to deliver sediments into the river mouth to form the north spit.

Therefore, in spite of the groynes, Coovam's mouth stayed, more or less, blocked throughout the year. It was hardly an improvement over the situation before the groynes, on the contrary, worse. A better solution may be to keep Coovam clean. Then, it would not matter if it drained fast enough to sea or not. The only saving grace was the Marina Beach. But the beach too may be under the threat of erosion.

From an unconfirmed source, it is understood that there are some serious plans to dredge the mouth of Adayar River, to make it navigable for large fishing boats or to create a harbour within the mouth. Adayar drains through a 'naturally' shallow mouth about six kilometres south of Coovam. The Marina Beach forms the northern half this stretch. The southern half is densely built up, with a road running very close to the narrow coast.

Dredging Adyar's mouth would amount to preventing the spit. That would cut off the sediment supply to the coast downstream, to the north. Without doubt, the coast north of Adayar will begin to erode as soon as the Southwest Monsoon sets in. To control erosion, the first-aid measure will invariably be a seawall. With no sediment supply to the downstream, the erosion will continue unabated north of the seawall. So they will extend the seawall further north. The erosion too will advance north. It will not be long before the erosion reaches the Marina Beach. The beach will begin to erode. Before long, an ugly seawall would take its place too. So the future of Marina Beach will become precarious, whenever the Adayar River Mouth is made spit-free by dredging.

CHAPTER 16

CANAL 'AT SEA'

We have seen how the coast gets its sediment supply. We have also seen how the sediments move on the coast, both along and across. Let us now examine what happens to the sediments that eventually reach the seabed, brought down by run-backs from the beaches or by runoff from the mudflats.

In both cases, it is clay rich fluid-mud that reaches the seabed. The mud, as it slowly slides downslope, tends to bunch together to form little balls of mud. The process is known as flocculation. What makes mud to flocculate on the seabed is still a mystery.

Let us leave it as a mystery and see if it serves any purpose. Nothing happens in nature without purpose. The purpose may even hold the key to understanding the mystery. The mud balls roll more easily down the seabed slope compared to the fluid-mud slithering down. The balls offer least resistance, hence require no esoteric underwater currents to cause them to move along the seabed. Gravity alone can handle the process.

The mud rolling down the seabed is an interminable process. As long as there is erodible material on land, mud will go on being delivered to the seabed across the coast, which then will go on rolling down all the way down to the deep sea bottom. Nature's aim is therefore to raze the land and fill the adjoining sea. That is all quite simple and straight forward. But what should interest us is what happens when the mud balls, rolling down the seabed slope under gravity, are impeded?

Obviously, the impediment, whether natural or manmade, will arrest the mud rolling down. Coastward of all nearshore islets and islands are therefore both shallower and muddier compared to their seaward. Likewise, mud would roll into channels dredged along the coast, as into a sand-trap. Therefore, a channel along the coast will silt. The port designers therefore align the approach channels as perpendicular to the coast as possible.

But two major ports on the East Coast, Chennai and Ennore, have their approach channels running along the coast. Chennai Port's channel heads north along the coast for about 2½ km before turning east into deep waters. Ennore Port's channel also runs nearly along the coast, but southward, till it hits deep waters.

Annual maintenance dredging at Chennai is about 0.36 million cubic metres. Nearly all that is dredged from the channel. Doubtless, most siltation must be happening in the portion that runs along the coast. From there sediments slip into the channel eastward, aided only by its seaward gradient. Even at Ennore, the maintenance dredging is almost entirely within the approach channel.

A channel perpendicular to the coast also will silt, but at a much slower rate, compared to one along the coast. Besides less siltation, a perpendicular channel need not be long, as deep water is quickly reached, because the seabed normally slopes seaward perpendicular to the coast. A perpendicular channel is also safer for the vessels. In a channel along the coast, the normally onshore winds can set a slow moving vessel coastward, into shallow waters.

What may have prompted the designers to align the approach channels of the two ports along the coast? Without doubt, it must have been their ardent belief in the theory of longshore currents. Anyway, both these channels are short. Siltation is being tackled by maintenance dredging, without undue adverse impact on profitability. Nevertheless, it still may be worth the effort of reorienting the channels to make them perpendicular to the coast.

There is now a long navigable canal being created along the coast, not one, but two, Indian and Srilankan. It is the Sethusamudram Ship Canal. It is intended for the seagoing vessels to sail through the Palk Bay. Its northern end passes through the shallow Palk Strait to enter the Bay of Bengal, and in the south, it cuts through the Adam's Bridge or Rama Sethu to enter the Gulf of Mannar.

At present, the vessels go around the island of Srilanka to go from Arabian Sea to Bay of Bengal or the other way. Sethusamudram Canal will shorten this distance by about 600 km. That is an advantage, but insignificant compared to the Suez Canal, which saves the vessels from rounding not a tiny island, but a continent, Africa. The proponents nevertheless call Sethusamudram the Suez of India.

There is however a fundamental difference between the two. Suez is an 'overland' canal, created by excavating land or passing through overland water bodies, whereas Sethusamudram runs entirely on the seabed. That is precisely the reason why the latter will not work.

The idea of a canal through Palk Bay was first mooted in 1860 by one Commander Alfred Dundas Taylor of the Royal Indian Marines. Taylor's proposal was a largely overland canal. But high cost did the idea in.

Over the next several decades, many others, both individuals and institutions, proposed similar overland canals, with small changes in routing. These too were not pursued due to high cost vis-à-vis advantages.

Post-independence proposals somehow became seabed based probably because the proponents felt dredging the seabed was simpler. Moreover, there was no need to acquire large tracts of land that an overland canal cannot do without. Again, due to lack of resources, the proposals were not followed up.

In the mean time, the dredging technology made great advances. From a port support activity, dredging grew into an independent industry. The growth was fuelled by large scale port development, worldwide, coupled with poor understanding of the coast. Everyone felt that a port could be created anywhere on the coast merely by constructing breakwaters and some dredging thereafter. Not only such ports silted, they created a host of problems on the coast, all of which needed more dredging to tackle. That spurred the growth of the industry.

The burgeoning dredging industry then began to scout for projects. Sethusamudram Canal that was in cold storage since long came as a bonanza for the industry, both for capital dredging, thereafter for maintenance dredging, probably forever! Several 'vested interests' soon came to be involved. They sought to have the canal project sanctioned, somehow.

Soon the canal was turned into an emotional and political issue. Rallies in support and protests against followed. It became a matter of people's pride with politicians, academicians and religious leaders coming out in support, but mostly orchestrated, hardly aware what the canal actually meant. For the vested interests, there was a lot of money to make, both licit and illicit. Costing in dredging is not quite a transparent process. Finally, bowing to the pressures of coalition politics the Central Government sanctioned the Sethusamudram Ship Canal Project.

The task of designing the canal and having it cleared environmentally was entrusted to the Nagpur based National Environmental Engineering Research Institute or NEERI, far away from any coast. NEERI not only lacked the nautical outlook, but also expertise and experience to design a major coastal project like Sethusamudram. But then, they could always hire the services of experts.

Come to think of it, there may be none with the requisite expertise and experience, anywhere in the world, because there are no such long seabed based manmade canals that run along the coast, anywhere. NEERI nevertheless came up with the design and routing of the canal, and also declared it environmentally safe. See the Figure 16.1 below. Capital dredging began in July 2005.

The canal is 168 km long, 300 m wide and 12 m deep. It is in three parts: Adam's Bridge, Palk Bay and Palk Strait. Adam's Bridge part is about 35 km long and requires about 48 million cubic metres of capital dredging. That also includes blasting through the natural rock ledge extending between India and Srilanka, the Adam's Bridge. Palk Bay part is about 79 km long. It has the required depths of 12 m and more. That is adequate for the first phase of the project. The proponents envisage a deeper canal later on. But for the present, it needs no capital dredging. Palk Strait part is about 54 km long, with about 35 million cubic metres of capital dredging.

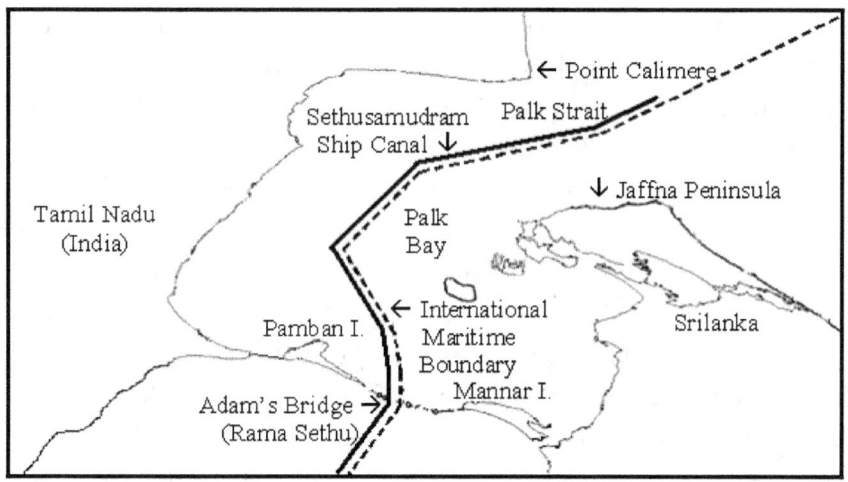

Figure 16.1: Sethusamudram Ship Canal

The estimated project cost in 2005 was ₹ 22.33 billion. Nearly 77% of that was for capital dredging—₹ 17.20 billion. The project cost has since been revised upwards, several times, quite substantially. Anyway, capital dredging is only a onetime cost. The recurring cost is for maintenance dredging. That will depend on the canal's rate of siltation.

NEERI estimated that the dredged depths would reduce by about ten centimetres in a span four years. In other words, the canal would silt at the rate of two and half centimetres per year. In terms of volume, that would be annually 0.55 million cubic metres.

The Final Detailed Project Report by L&T-Ramboll Consulting Engineers Limited, Chennai, put the siltation at two million cubic metres per annum. The report, in addition, stated that the siltation would reduce to 1.4 million cubic metres per annum after about five years. They however did not foresee any changes in the wave climate or rainfall pattern in the region that may have some bearing on sediment movement, hence the siltation. Which of the two figures is correct? Both cannot be. Yet both can be wrong!

To get an idea of siltation in Palk Strait stretch of the canal, we can take the siltation in the channel of Chennai Port as a yardstick. The channel silts about 0.36 million cubic metres per annum. Though the canal is about 6.7 km long, siltation takes place mainly in the 2½ km stretch that runs along the coast. Therefore, the annual rate of siltation, with coast only to one side, can be taken as 0.12 million cubic metres per kilometre.

Rainfall and wave climate on the Indian side of Palk Strait are comparable to that at Chennai. Besides, there are several seasonal streams that discharge sediment-laden waters into the strait. In addition, there is a significant longshore drift southward during Northeast Monsoon from the East Coast that delivers sediments into the strait. Therefore, sediments migrating into the Palk Strait from north would be more than what goes into the channel of Chennai Port. Let us therefore peg the sediment supply into the Palk Strait stretch from north at a modest 0.15 million cubic metres per kilometre.

Sediments also migrate into the strait from the south, from the Jaffna Peninsula. That may be more than that from the north, because the peninsula gets more rainfall, therefore sends more sediment-laden runoff into the strait. In addition, there is a much stronger longshore drift during the Southwest Monsoon from the east coast of the peninsula delivering sediments into the strait. Again, as a very modest estimate, let us peg the sediment supply into the Palk Strait stretch from south also at 0.15 million cubic metres per kilometre.

Therefore, the stretch would silt annually at a rate of 0.3 million cubic metres per kilometre. With a length of 54 km, the annual siltation would be 16 million cubic metres. We can actually expect the stretch to silt up fully every year, which means 35 million cubic metres of siltation per year. But for the present, let us stick to the highly modest figure of 16 million cubic metres per annum. That is 8 times the estimate made by L&T-Ramboll and 30 times that of NEERI!

Annual maintenance dredging cost, as per 2005 rates, would be ₹ 1.6 billion, whereas L&T-Ramboll's estimate then was only ₹ 200 million. If we go by current rate, it would be lot more. And in the future, it is anybody's guess. That is an annually recurring cost!

Besides the recurring cost, where would they find dredgers to dredge such a huge quantity, year after year? The canal may be plying only dredgers, instead of the cargo vessels. How then will it generate revenue? What about the environmental impact of dumping so much dredged material, year after year? What about the maintenance dredging of ports and naval bases, if all the dredgers are engaged in maintaining the canal? Will the dredging companies invest in more dredgers, only to dredge the canal that may not be able pay them back? There are too many difficult questions, but no answers.

The canal through Adam's Bridge is in three sections. North section, within Palk Bay, is about 20 km long. The section spanning the rock ledge is about 4 km long. The one south, within Gulf of Mannar, is about 11 km long. Unlike the Palk Strait stretch, the canal here runs almost perpendicular to depth contours, similar to a port channel perpendicular to the coast. Moreover, there is no sediment source around. Therefore, siltation is not an issue in this portion.

But there is another problem that would elude solution. Strong currents flow through the Pamban Pass, north of Adam's Bridge, and also through the numerous narrow passes within the Adam's Bridge, all year round, one way or the other. During the Southwest Monsoon, the currents are northerly, flowing from the Gulf of Mannar into the Palk Bay. During the Northeast Monsoon, the currents flow the other way.

The reason for these currents is not anything esoteric. We have already dealt with the phenomenon. It is due to the storm surge. During the Southwest Monsoon, the strong south-westerly winds generate the storm surge south of the Adam's Bridge. Because of the raised sea level south, water literally cascades north through the passes. During the Northeast Monsoon, the surge develops north, though not as high as on the south. The resulting southerly currents are therefore weaker.

Currents flowing through the Pamban Pass are vicious for small craft navigation. Accidents are frequent. Only an experienced helmsman can take the craft across safely. The currents flowing through the passes within the Adam's Bridge too are swift and turbulent, but not of any consequence, because the passes presently are not navigable. But after the Adam's Bridge is blasted through, the currents will continue to flow, one way or the other, depending on the monsoon, just as swiftly and turbulently. That would make the navigation of the slow-moving vessels both difficult and dangerous.

Work on the canal has been since suspended, but after spending billions on capital dredging. The reason for suspension is neither scientific nor economical, but religious. The Hindus consider the Adam's Bridge, or as they call it the Rama Sethu or Rama's Bridge, sacred. They believe that Sri Rama, the legendary King of Ayodhya, built the bridge to go to Srilanka, to rescue his kidnapped spouse. Therefore they oppose the blasting. Without blasting, the canal cannot be.

The Palk Strait stretch, where most of the capital dredging has been done, beyond any doubt, must have silted back fully, leaving hardly a trace of the dredging. The Sethusamudram Canal for all practical purposes is finished. Of course, the loss would be borne by the unsuspecting taxpayers. Anyway, that may be lot better than the recurring cost of maintenance dredging, if the canal would have been commissioned.

There are more siltation stories. The first major case of siltation, though not publically reported or discussed, happened in 1964, at Karanja, within Mumbai Harbour. It was Navy's first encounter with siltation. We have already noted the second one. That happened in 1979, the siltation in Naval Dockyard Mumbai.

CHAPTER 17

KARANJA DEBACLE

Navy's armament depot was on the Butcher Island, a small island north of Mumbai Harbour. See the Figure 17.1 below. In 1952, Bombay Port Trust (BPT) sought the island to build an oil terminal. That later became the Butcher Island Oil Terminal. Navy was given land at Karanja, about eight kilometres east of the Naval Dockyard, across the port's channel, to build a new armament depot.

The construction of the depot began in 1953. The Naval Armament Depot Karanja was ready by 1959. Transporting ammunition to warships berthed at the dockyard in ammunition barges, across a busy shipping channel, or in ammunition trucks, through Mumbai's crowded traffic over a distance of 40 km one way, proved risky, laborious and huge waste of time. So Navy proposed an ammunitioning base at Karanja, where the warships could go alongside to safely and quickly embark or disembark ammunition. The Central Government approved the proposal. CWPRS was tasked to design the base within the Karanja Bay.

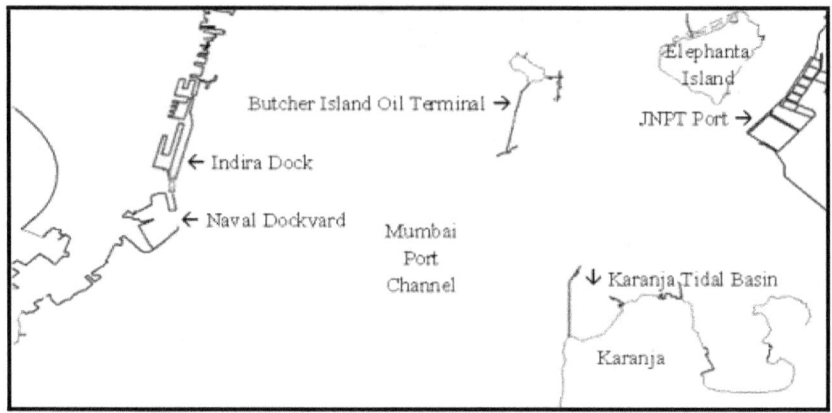

Figure 17.1: Butcher Island to Karanja

The bay was in the lee of Karanja Hill, hence naturally tranquil throughout the year, even during the monsoon. The bay's shoreline was rocky. From the shoreline, a shallow reef extended offshore about four to six hundred metres. Beyond the reef, the bay was quite deep. A survey by INS Bundelkhand, an old surveying ship, put the average depth in the bay at about four fathoms, that is, about eight metres!

The Naval Dockyard Expansion Scheme or NDES required huge quantities of rocks and boulders. The quarry was on the Karanja Hill, next to the bay. NDES contractors, the PIM Construction Company, built a jetty on the bay's eastern shoreline to embark the rocks and boulders into the barges for transporting to the dockyard. The jetty therefore came to be called the PIM Jetty. Some old timers claim INS Mysore, the old cruiser of World War II vintage, was once berthed at the PIM Jetty for de-ammunitioning. That was only to stress that the bay was deep, even along the jetty.

The Karanja Bay was deep and tranquil, right next to the armament depot, therefore an ideal site for an ammunitioning base. Nothing much was needed to turn it into one. A T-headed jetty on the bay's eastern shoreline should have sufficed. No one thought of that. Instead, CWPRS came up with the design for an enclosed tidal basin, the Karanja Tidal basin, enclosed by the Karanja Breakwater. See the Figure 17.2 below.

A breakwater is usually built to create a tranquil zone within a bay that is open to waves. Did the designers not realise that the Karanja Bay was naturally tranquil? They worked on the tidal model of Bombay Harbour. The tidal model is not suitable for assessing tranquillity, either existence or lack. A wave model is needed for that. Actually, there is hardly need even for that. A site visit would have made it quite obvious.

Probably, the designers opted for the breakwater, not so much for tranquillity, but to prevent the littoral currents delivering sediments into the bay. According to their theory, rather dogma, the littoral currents delivered sediments from sea into the harbour. The same dogma was behind the redesigning of NDES only a couple of years ago. If the bay had not silted until then, where was the need for a breakwater? No one asked that question. Dogmas however resist change, even in the face of overwhelming counterevidence.

The counterevidence had come in 1959, barely few months before they got down to designing the Karanja Tidal Basin. It may have been quite fresh on their minds. It was one of the major field-studies that CWPRS undertook at Mumbai, the radioactive tracer study, the first of its kind anywhere.

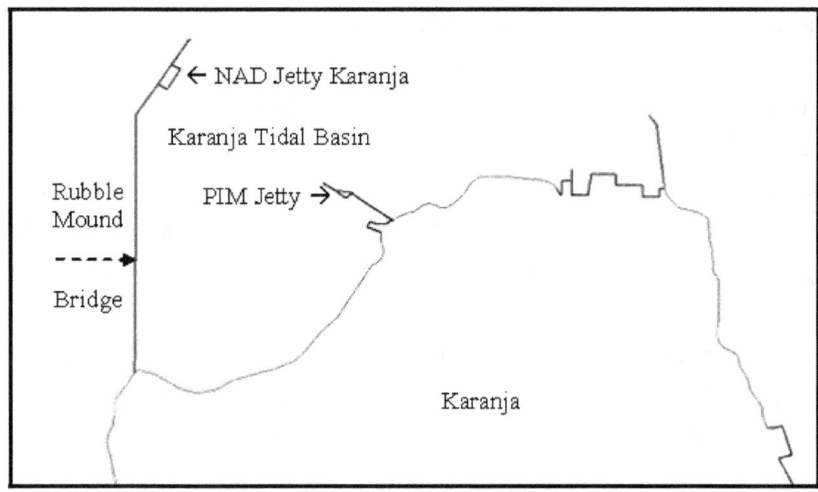

Figure 17.2: Karanja Tidal Basin

The construction of Butcher Island Oil Terminal involved extensive capital dredging. They needed a place to dump the dredged material, but not too far from the construction site. A distant dumping site would make the dredging expensive. On the other hand, if it was too close, they believed that the dredged material would be transported into the harbour by the littoral currents to cause siltation. Therefore, the dumping site had to be carefully located. That led to the radioactive tracer study.

The study began in March 1959. The Hydraulic Research Station UK, the Indian Atomic Energy Establishment and the BPT assisted CWPRS in the study. They used tracers made of finely ground Boron-free glass carriers, bearing Scandium Oxide, with the radioactive Scandium 46 (half-life period – 85 days), to simulate sediments in suspension.

The tracers were scattered at the harbour mouth and continuously tracked using special detectors. Tracking went on till 03 May 1959, when the worsening sea conditions forced them to call off the study. Track charts were prepared on a day-to-day basis, which after analysis showed that no tracers had in fact entered the harbour with littoral currents or by any other means. Therefore, a dumping site just off the harbour mouth was indeed safe.

While the radioactive tracer study solved the dumping site problem, it did little to alter the dogma about littoral currents. Karanja Breakwater is a testimony that the dogma still held sway.

Karanja Breakwater is 1,800 m long, in two sections, a 1,200 m long rubble-mound and 600 m long bridge. Rubble-mound is made up of large boulders laid systematically on the seabed, in layers, with a concrete topping to serve as road. Rail tracks are also embedded on the concrete topping. The 120 m long ammunitioning wharf, the NAD Jetty Karanja, is on the rubble-mound, towards the northern end. See the Figure 17.2 above. The bridge stands on several smooth round concrete piles, with its southern end resting on land at the southwest corner of the bay and the northern end on the rubble-mound.

The breakwater's construction began in 1960. It was ready by 1964. When Navy took over the tidal basin, they found it silted, completely. Many attempts were made thereafter to de-silt the basin, but to no avail. The basin became a mudflat. It could no longer be used as an ammunitioning base—an apparently unredeemable case. So the Karanja file was closed with no one getting any wiser. Incidentally, the project cost the nation ₹ 1.6 billion, a princely sum in the 1960s.

The ammunition continues to be delivered to the dockyard in barges and trucks. NAD Jetty Karanja is now used for berthing ammunition barges and passenger ferries that ply between the dockyard and Karanja. Even that requires regular maintenance dredging, which can be done only by small grab-dredgers.

CHAPTER 18

SEISMIC SURVEY

My first encounter with the silted Karanja Tidal basin took place in 1983. I was then aboard INS Darshak, as a junior hydrographic surveyor. One morning, the Commanding Officer asked me to rush to Karanja to render some surveying assistance to a team from CWPRS. He did not give me any detail. He probably had none. I left immediately with two survey motor boats, Spica and Polaris, with full crew and gear, to be on the safe side, not knowing the nature of work.

The CWPRS team was waiting on the jetty. They appeared to be in a hurry. They had arrived in a truck that resembled a mobile laboratory. They were there to do the seismic survey of Karanja Tidal Basin. I was there to provide the platform, the survey motor boat, and necessary surveying support. DGNP (B) was coordinating.

We loaded their equipment into Spica. Polaris was kept standby. Besides the seismic survey system, there was a Honda generator for power supply. We lashed the generator on the boat's bow deck, right in the front. That would not have been a right thing to do, if we had to go out to sea. Sea spray would hit the generator. But on that day, we were only operating only within the basin, where it was tranquil. Actually, there was no other place for the generator. Rest of their equipment had completely occupied the boat's rear cabin.

With all things loaded, and with the CWPRS team and boat crew aboard, Spica was down to gunwale, the upper edge of the hull. If the boat rocked, even slightly, it would ship in water and sink. The situation was precarious. Keeping only bowman and engine driver, I disembarked rest of the crew, including coxswain. I took over the helm. Bowman was needed to keep a lookout ahead and to operate the generator. Also, the boat cannot run without the engine driver. Besides three of us, there were three from the CWPRS. The situation still remained precarious.

I gave clear instructions to everyone onboard not to move about without my express clearance, and any unavoidable movement, person or equipment, had to be done gently. Sudden shifting of weight would send the boat rocking. Everyone had life jacket on. I reiterated the risks. Everyone, I hoped, understood. I castoff and headed to slightly deeper waters.

The team leader then suggested lowering the acoustic thumper over the side. He did that himself, very carefully, without rocking the boat. The thumper is a heavy transducer. It is the sound source for the seismic surveying system. It produces powerful blasts of ultrasonic sound, which is inaudible to humans. But if you are underwater nearby, it would shatter your ear drums and probably other internal organs too.

Thereafter, on instructions from the leader, I streamed the hydrophone array across the boat's stern. Hydrophone array is a cable stringed with hydrophones. It was about 100 m long. Hydrophone is the device to listen to the ultrasonic sound underwater.

The system works like this. The acoustic thumper sends down a powerful blast of ultrasonic sound into the seabed. Some energy gets reflected back from the seabed. That gives the depth of water below the thumper. Rest goes down into the seabed, which after refractions through layers of mud and rock returns to the hydrophones at different intervals. The speed of sound depends on the density of medium. Layers of mud and rock below the seabed had different densities. The system, after processing the refracted echoes received by the hydrophones, gives depth, density and thickness of mud and rock layers below the seabed.

Deep sea seismic surveying systems aboard seagoing seismic surveying vessels use massive air guns as the sound source. Their hydrophone arrays normally stretch several kilometres behind the vessels. Deep sea seismic survey is done to study the structure of earth's crust below the seabed, primarily for offshore oil explorations.

At Karanja, they were only trying to measure the bedrock depth, in other words, the thickness of mud over the bedrock, using a shallow water seismic surveying system.

With the thumper over the side and hydrophone array streaming astern, the leader decided to turn on the system. I asked the bowman to start the generator. A press of the button did it. The purr of the generator was barely audible over the boat's engine noise. I was impressed. I was used to the noisy hand-cranked diesel generators that we used in our survey camps.

After the generator came on, the leader waited few minutes for the supply to stabilise. He carried out some more checks and turned on the system. Few seconds later, there was smell of burning. The processor started to smoke. Without waiting for any instructions, I ordered the bowman to shut down the generator by hitting the red knob, the emergency shutdown button. I also ordered shutting down boat's engine. Engine driver then stood by with the fire extinguisher. Tense moments, the smoke persisted for a while. Fortunately, there was no fire. Though it was a shallow water seismic surveying system, it was probably not suited for areas as shallow as the silted basin. Return signals were too strong for the system to withstand. We returned to the jetty after carefully recovering the thumper and array. That was the end of seismic survey, at least, till an alternate was found.

I got orders not to return to ship, but to stay put at Karanja. We were billeted at INS Tunir, a technical shore establishment at Karanja. I learnt from the leader that their system would take a long time to be made operational. The spares had to be imported. In any case, they could not make another attempt in the shallow basin with a similar system. They had to find some other way to measure the bedrock depth.

In their laboratory truck, there was another seismic surveying system, but for use on land. The system consisted of a ground-phone array, similar to the hydrophone array, connected to a recorder-cum-processor. There was no noise generator, like the acoustic thumper of the marine version. The noise was generated by setting off an explosive at a specified distance from the ground-phone array. The array was laid on the ground in a straight line, with each ground-phone buried few centimetres in the soil. I felt that the system could be deployed in the silted basin. But we needed the help of divers and underwater explosives.

On the following morning, I briefed the DGNP (B), the Rear Admiral who headed the organisation. He had flown down to Karanja in a helicopter. Divers and explosives were quickly arranged from INS Abhimanyu II, which was then the training centre for clearance divers, also situated at Karanja, right next to the basin. Clearance divers were experts in the use of underwater explosives. One of their tasks was to de-fuse the underwater mines laid in ports and those attached to ships' bottoms, the limpet mines. They were exceptionally proficient underwater, even in the slushy soft mud in the basin. After all, it was their training ground.

Divers laid the ground-phone array on the basin's bed in a straight line exactly where we wanted. They also laid the explosive charge, known as the depth-charge, at the required distance from the array. I fixed the positions of ground-phone array and depth-charge with a sextant. We had earlier put up marks, bamboos with flags, on the breakwater for position fixing.

CWPRS team was on the breakwater with the recorder-cum-processor. At a signal from the team, divers detonated the depth-charge. Echoes came back to the ground-phones after refraction from the various layers below the bed, just as in case of the marine version. The team was satisfied with the results. We repeated the procedure along several lines across the basin. It took us about a week. The task was laborious and risky. It had to be carefully executed. Divers took care of the safety with precision.

109

After preliminary processing, the team found the bedrock was at a depth of about fourteen metres. In other words, there was fourteen metres of mud over the bedrock! But somehow, I did not ask them why they were interested in the bedrock depth. It was only much later, during my tenure at the ASD (B) Survey Unit, while going through some of the old files, did I learn that they were looking for a way to desilt the basin. They were hoping to cut a channel through the bedrock to drain the mud that had collected in the basin. Drawings that accompanied the report showed several orientations of the channel. But there was no mention how they were going to cut such a channel on the bedrock. Probably, they awaited the results of seismic survey.

They probably did not expect the result, the bedrock fourteen metres below. To cut a channel through the bedrock, they must first dredge enormous amount of mud. That by no means would be easy. As they dredge, the surrounding mud would slip in, so the dredging would go on for long time and cost a lot too. Only after that could they attempt to blast the channel through the bedrock. That again would involve enormous effort and cost. Moreover, all that must be completed before the onset of next monsoon!

During the seismic survey, I did however get to know how CWPRS designed the Karanja Breakwater, and why, according to them, the basin had silted.

CHAPTER 19

BRIDGING THE RUBBLE-MOUND

Designing the Karanja Breakwater happened in three stages. The designers started out with a 1,800 m long straight rubble-mound breakwater. It stretched north from the southwest corner of the Karanja Bay, with the basin forming on its eastern side. So the basin's entrance was from the north. Vessels entering or leaving the basin would therefore have to do a 180 degree turn around the breakwater's northern tip. That would have been quite cumbersome. It is unlikely that the designers would have given much thought to that. They were not mariners.

They built the model of breakwater, to scale, and placed it on the tidal model of Bombay Harbour, the one they had earlier used to redesign the NDES. Then they simulated the tides. During the ebb tide, they observed turbulent eddies around the breakwater's northern tip. On the model, that must have been quite spectacular, because the 'simulated' ebb stream flowed about ten times faster. The model was vertically exaggerated ten times.

Turbulent eddies would pose problems for the vessels already quite encumbered while entering or leaving the basin. It is difficult to handle vessels through eddies. The steerage is sluggish. Again, the vessel handling may not have been the designers' concern. Eddies also meant the ebb stream around the breakwater's tip was slowing down. That bothered them. They believed that the tidal stream, both ebb and flood, bore sediments in suspension. If the sediment-laden ebb stream slowed down, somehow, the suspended sediments would settle down to the bed. The area would silt. So they had to prevent eddies.

In the second stage of designing, they gave the straight rubble-mound breakwater a 30° eastward bend, at around 450 m from the tip. That, in fact, aligned the bent portion with the 'simulated' ebb stream. Then they tested the bent breakwater on the model. There were no more eddies.

Thereafter, they had to ensure water in the basin kept flowing continuously, so that the suspended sediments would continue to stay in suspension. If the water stopped flowing, suspended sediments would settle down to the bed. That was how they understood the process of siltation.

They sprinkled 'treated' sawdust on the model to simulate sediments in suspension. They had no other means to simulate that. The sawdust was treated by rinsing it in saturated lime water. After washing it with freshwater, it was dipped into one-percent solution of copper sulphate. The purpose of the treatment was to ensure the sawdust did not get soggy and sink, at least, for the duration of study. Today, they probably would use bits of polystyrene, instead of treating sawdust.

Treated sawdust floated on the model, and remained floating always, no matter the water flowed or not. A body held in suspension behaves quite differently. It would settle down to the bed, when the water slowed down or stopped flowing. On the other hand, the floating body would keep moving as long as the water on the surface kept flowing. It would only stop moving, when on the surface the water stopped flowing.

So wherever on the model, the sawdust stopped moving or gathered up in one place, the inference should have been no water flowing there on the surface. But the designers deduced that as stagnation of water. When the water stagnates, siltation would follow, because the suspended sediments would settle down to the bed. That was the theory.

They sprinkled the treated sawdust at the north of the model. The sawdust floated into the model basin, east of the breakwater, along with the simulated ebb stream. Whatever sawdust entered the basin stayed put. It did not flow out even with the change of tide. So they concluded water in the basin was stagnating. Does that happen at the coast?

At the coast, during the rising tide, the water level rises everywhere. To make that happen, water flows in as flood stream. When the tide falls, water then flows out to sea as ebb stream to bring the level down, everywhere. Therefore, at the coast, water is flowing continuously, one way or the other. Probably, for a short duration, around high or low tide, when the flow reversed, it may seem a bit like stagnation.

Flotsam or floating debris that enters a tidal basin with the flood stream invariably flows out with the ebb stream, unless it is trapped in some nook or cranny. Sawdust floating on the model is no different from the flotsam. It should have flowed out, provided the model simulated tidal stream correctly, like it is at the coast.

Tidal stream is a surface phenomenon. It flows only above the Chart Datum, hence remains unaffected by the bed gradient. But on the model, the simulated stream is powered by gravity. Hence its rate depended on the gradient of the model's bed, as in the case of a river. In a river, water flows along the bed at a rate dictated by its gradient. Unlike the surface flowing tidal stream, which diminished with depth, the river's flow increased with depth. In other words, in a river rate of flow is always more along the bed than at surface. Where the riverbed has a shallow gradient, the surface flow may be hardly apparent, even though water continues to flow downslope along the bed. That probably was the situation in the model basin. The surface flow on the model basin may have been too weak to move the sawdust. Therefore, sawdust staying put in the basin was the model's limitation than the representation of any natural tidal process.

But the designers did not think so. They deduced that water in the basin was indeed stagnating. If so, the suspended sediments would settle down to bed. The basin would silt. But that would have happened only if there were sediments suspended in water. That could have been easily ascertained by collecting water samples from Karanja Bay and filtering out a measured volume of each sample. Had they done so they would have realised the water bore no sediments in suspension, not only in the bay, but entire Mumbai Harbour.

Since they had assumed water in the basin was stagnating, they had to get it flowing again. That led to yet another modification, and a very expensive one at that. Instead of the breakwater remaining entirely rubble-mound, they converted a 600 m portion at the southern end into a bridge. But that put the bridge entirely on the shallow reef that fringed the bay's inner shoreline. Water would therefore flow beneath the bridge only for a short duration around the high tide. Did the designers realise that?

The tidal model they worked on was built from the chart of Mumbai Harbour, Chart 2016. This chart, like any other, did not accurately represent the shallow non-navigable areas along the coast, particularly the reef. Actually, it is extremely difficult to survey the reef. Survey motor boats cannot go there. Even the rubber dinghy would not work. Sharp edges of the reef would shred the dinghy. It is also risky to go on foot, with a sounding pole. But then, where was the need to take so much trouble to survey a reef along the coast? No vessel would venture there anyway. Normally, only the reef's outer limit is shown on the chart. Even that is only an estimate made by the hydrographer as his survey motor boat nears the reef, at the end of each sounding line. He would record that in his sounding book, for example, 'reef ahead – 5 m'.

As a result, there was little detail of the reef on the chart, but for its rough outer limit marked by a jagged black line. The reef itself was shown in a uniform shade of olive green. The first depth contour along the reef on the old Chart 2016 may have been the one-fathom line, that is, nearly the two-metre contour. Those who built the model were no experts in interpreting charts. They probably assumed the gradient of coast was smooth and gentle from the charted high water line down into water up to the one-fathom contour. In other words, the reef did not show up on the model.

When they tested the bridge cum rubble-mound breakwater on the model, the simulated tidal stream, ebb and flood, flowed freely through the bridge. No sawdust was trapped in the basin! Therefore, with the bridge in place, they concluded, the water in the basin would not stagnate. If water did not stagnate, the suspended sediments would not settle down. The basin would not silt. What about the littoral currents delivering sediments through the bridge? Probably, no one thought of that. Not that it mattered. There are no littoral currents capable of moving sediments anyway.

Soon after the design was ready, the go-ahead was given for the construction. But the contractor faced a huge problem. The rubble-mound was not connected to land. Therefore, the boulders would have to be ferried in barges. A floating crane would also be needed to hoist the boulders from barges to lay the rubble-mound. That would make construction cumbersome, slow and expensive.

On the other hand, if they built the bridge first and used it to deliver boulders to the rubble-mound, in trucks, then by the time construction would be completed, the bridge would be wrecked too. Both options therefore did not look good.

So they came up with a novel solution. They proposed building a temporary motorable track across the reef, where the bridge was slated to come up. They would then use the track to ferry the boulders in trucks to build the rubble-mound first. After completing the rubble-mound, they would excavate the track, down to the original bed level, and build the bridge in place.

Apparently, no one had objections to the proposal, most probably, not even the CWPRS, assuming that they were monitoring the execution of their design, as is the customary practice. It is also quite unlikely that such a major decision on the project execution would be taken without involving designers. Most probably, everyone, at that time, must have felt it was a good solution. And, it was! It saved time, effort and money.

The contractor built the track using rock debris from Karanja Quarry. The track made construction of the rubble-mound simple and quick. After that got over, they excavated the rock debris down to the original reef bed and built the bridge in place, exactly as per the design. So by 1964, Karanja Breakwater was ready. We know what happened after that—a completely silted Karanja Tidal Basin!

CWPRS investigated the matter. They held the contractor guilty. The temporary motorable track, according to them, prevented the sediment-laden tidal stream flowing freely through, as their model had shown. That, they concluded, led to the stagnation of water in the basin, hence the suspended sediments settled down to the bed and so the basin silted. That was a simple argument. No one at that time challenged it.

To silt up the basin with an area of about one and half million square metres and average depth of eight metres needed at least twelve million cubic metres of mud! There were no sediments suspended in the tidal waters that flowed through the basin. But the siltation was real indeed. In other words, twelve million cubic metres of mud did fill the basin!

Where did so much mud come from and how?

CHAPTER 20

MUMBAI'S MUDFLATS

The siltation in the tidal basins of dockyard and Karanja did not quite agree with the general understanding of the phenomenon. There are no sediments suspended in the water anywhere in the harbour. But on the bed is a layer of fluid-mud, thick creamy mud, too dense to be held in suspension. If this mud moves, it can only do so along the bed. It cannot exist in the slow surface-flowing tidal stream.

It could not have come from sea either. Nothing so dense can flow upslope, against gravity. Also, it could not have come with littoral currents, even if there were such currents capable of moving dense sediments, because the breakwaters, South and Karanja, were specifically intended to prevent that.

Therefore, logically, the mud must originate within the harbour, upstream of the basins. Where in the harbour can we find so much mud? Actually, it is everywhere along the harbour's inner shoreline, but predominantly at the mudflats—Sewri, Nhava-Sheva, Uran and others. See the Figure 20.1 below.

Uran Mudflat is close to the Karanja Tidal Basin. Therefore, we can infer that the basin got its mud supply from Uran. Likewise, the dockyard got its supply from the Sewri Mudflat, the nearest one upstream.

We can confirm that Sewri is the dockyard's mud source by a radioactive tracer study. Spread some tracers on the mudflat and track them. Some tracers will surely find their way into the dockyard, but not in every season. The mud flowed out of mudflats only during monsoon. The inference is based on the study of siltation in the dockyard. The dockyard silted only during the monsoon, not before and not after.

The tracer study may not work at Karanja. Since the basin is fully silted and not dredged, no fresh mud would flow into it. Therefore, the tracers would not enter the basin. But if dredged, like the dockyard, the mud, along with the tracers, would flow into the basin immediately with the onset of monsoon.

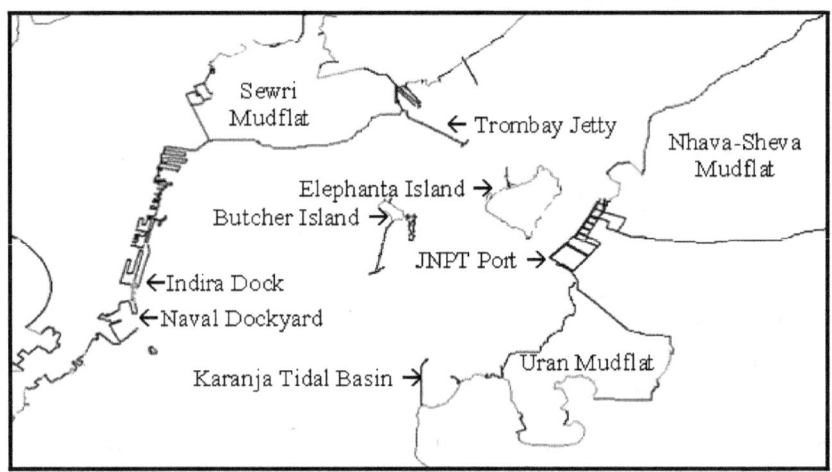

Figure 20.1: Mumbai's Mudflats

But how does the mud get flowing from the mudflats? It is a monsoon phenomenon, therefore linked to the rains. The dockyard silts as soon as it begins to rain. That is a good cue to infer the process. Mud is always present at the mudflats, but during non-monsoon months, it stays congealed, hence too viscous to flow out. As it begins to rain, the runoff flows copiously into the mudflats. That in turn fluidises the viscous mud into a flow-able form. Almost immediately, the mud begins to flow out.

The outflow would stop only when the runoff ceases to flow in. That happens when the monsoon ends or takes a long break. Runoff also brings in fresh mud into the mudflats. That makes up for what is flowing out, hence the mudflats remain quite stable, year after year.

How does the mud flow further downstream after starting out from the mudflats? It moves along the bed, downslope, aided only by gravity. It may seem like the bed itself moving. The rate therefore is dictated by the bed gradient only. Because it is flowing along the bed, it remains unaffected by the surface flowing tidal stream. No matter the direction of tidal stream, the mud flows on seaward, as long as the runoff flows into the mudflats.

Being slow moving and under gravity, at no stage of its journey seaward, can the mud ride over any obstruction, natural or manmade, that stands above the level it is flowing. When the flowing mud comes against an obstruction, it does not stop flowing, instead skirts the obstruction, one side or the other or both, and continues to flow seaward. But when it encounters a deep, but semi-enclosed basin that opens into the flow, like the dockyard or Karanja, it flows in, as if into a sand-trap. But that does not go on indefinitely, only till the basin fills up to the level of incoming mud. That takes little time. Recall how rapidly the dockyard silted as soon as the rains started. Thereafter, nothing flows in. The mud then skirts the basin and flows on seaward, as if the basin did not exist.

But after a while, the mud that flowed into the basin congeals, and sinks a little. That makes room for some more mud to flow in. The process is repeated several times during each monsoon, till the mud is quite compacted.

The mud enters the dockyard, only because its mouth faced north, into the inflow. For this very reason, we can say with certainty that the Gibbs' dockyard would not have silted. See the Figure 7.2 above. Obviously, they knew about the mud flowing out of Sewri Mudflat during monsoon. That is evident from the design of Alexandra Dock, nee Indira Dock. The dock, built in 1914, has not silted, only because the mud cannot flow into it. The mouth faces south in the same direction mud is flowing. Mud flowing under gravity cannot do a right-about turn to enter the dock.

Therefore, turning the dockyard southward would also make it non-silting. There may be many ways of doing that, but the ideal way would be to restore the Gibbs' design. That is not such a difficult thing to achieve. Then, besides getting a roomier and siltation free dockyard, the graving drydock can be built as originally envisaged!

The Karanja Tidal Basin, being very close to the Uran Mudflat, silted to nearly the mudflat's level. But that was a shade lower than the level of the reef that fringed the basin's inner shoreline. Therefore, no mud from the mudflat flowed over the reef even before the Karanja Breakwater was built, and also later, from the silted basin below the bridge that was built entirely over the reef. In other words, the bridge had no impact on the siltation in the basin!

And for the same reason, the temporary motorable track built over the reef could not have caused the siltation! The basin, in fact, silted in stages, as the rubble-mound was built, every monsoon that ensued. The rubble-mound trapped the mud upstream. But no one took notice of that during the construction. The basin silted completely during the monsoon that followed completion of construction of the rubble-mound, long before the bridge was built.

The basin can silt no more. But if dredged, surely it will silt up, fully, in the next monsoon. That makes the Karanja Tidal Basin in the present form such a no-go case. But now that we know the process, we can use it to advantage to turn the basin into a non-silting one. Such a basin would be much larger than the dockyard, with a possible operating depth of 14 m, without any rock blasting. In fact, it can be turned into an ideal naval base, where even the aircraft carrier Vikramadithya, ex-Admiral Gorchakov, currently being refurbished in Russia, can be berthed. At present, there is no base where this carrier can be safely berthed.

INS Kadamba, the new naval base south of Karwar, where the construction of wharf and allied facilities are believed to be in progress, may not be safe.

Fundamentally, the base lacks shelter from the strong and often gusting monsoon winds. The aircraft carrier, with high freeboard, stands exposed to winds. Freeboard is the portion of a vessel above the water, hence affected by winds. Berthing a high freeboard vessel alongside in stormy conditions can mean inviting trouble.

When storm is expected at a port without wind shelter, for example, Chennai or Ennore on the East Coast, high freeboard vessels are ordered to sail out. Such vessels are in fact safer at sea than in a shelter-less port. Merchant vessels can be sailed out in the event of a storm, with little notice. Moreover, on the East Coast, the storms are less frequent, whereas on the West Coast, it is stormy conditions right through the four-month long monsoon, that is, for a third of the year! Where will the carrier then go? What if it is on refit and unable to sail, when an unseasonal storm hits? Therefore, Karanja may be the only choice.

To make Karanja non-silting, all we need do is to prevent the mud entering after the basin is dredged. A breakwater across the basin's present north-facing mouth would arrest the inflow. It then can be dredged. Thereafter, it is only a matter of creating a south-facing entrance to the basin, between this breakwater and the existing Karanja Breakwater, with an approach channel to connect the basin to the port's main channel.

In 1997, I had proposed the design for such a non-silting basin at Karanja—a Vision 2020 Paper. I learnt later that the Naval Headquarters had constituted a committee to study my proposal. But it was trashed, apparently, on a rather simple remark by a senior hydrographer—*"Yet another base in Mumbai will be like putting all the eggs in one basket."* That however did carry weight, because of the huge investments being made to create the third naval base, south of Karwar—INS Kadamba.

Few months after the incident, *all the eggs had to be put into one basket*. The irony was not lost on me. That happened when India came literally to war during the Kargil imbroglio. Navy was mobilised. Operational ships of the Eastern Fleet were sailed to Mumbai and placed under the command of the Fleet Commander Western Fleet.

Having spent some time in the dockyard, I could easily visualise the chaos that would have ensued. With just one fleet, the dockyard was congested and almost chaotic. With two fleets jostling for space, it may have been a crisis.

Navy needs more operating room at Mumbai. There is no more evading that. Easing the congestion by basing warships elsewhere may not be the solution. That would bring down the much needed force-level at Mumbai. That may be neither strategically nor tactically prudent. Mumbai and the surrounding seas, with huge offshore and onshore assets, will continue to remain vulnerable, always. In a crisis, INS Kadamba may be too far.

With so many agencies vying for waterfront space at Mumbai, Navy's need for additional operating room can only be met at Karanja. It is, in fact, tailor made. The base could then be used exclusively for berthing the operational warships and submarines. That would free the dockyard for the vessels on refit. That would also make the dockyard more efficient and productive.

CHAPTER 21

RIVER THROUGH THE LAGOON

New Mangalore Port silts every year, during the monsoon, just like Naval Dockyard Mumbai, but with hard mud, in fact, clay. After every monsoon, about six million cubic metres of clay is dredged from the port and its approach channel. The maintenance dredging accounts for nearly quarter of the port's operating cost!

Come to think of it, the port must not silt at all. It is neither in a creek nor in an estuary. A port inside a creek or estuary can silt, because the sediments flow through, quite naturally, particularly during the monsoon. New Mangalore Port is inside an artificial lagoon, created entirely by excavating into land. Therefore, no water ordinarily flows through the port to sea, with or without sediments. Only the surface flowing tidal stream flows in and out, which at Mangalore is weak, because of the small tide range. Yet the port silts profusely during the monsoon. That means sediments are somehow finding their way into the port.

The port's entrance is a 450 m wide artificial opening on the coast, hence open to sea. Waves would have progressed through this opening into the lagoon to disturb the tranquillity within. But that is not happening. It is quite tranquil inside, even during the monsoon.

After studies on a wave model, the designers found that a pair of breakwaters, one on each side of the entrance, reduced the intensity of waves progressing into the lagoon. The two breakwaters refracted the waves. Because of their length, position and inclination, the waves refracted by one 'destructively' interfered with those refracted by the other. That significantly diffused the waves at the entrance, creating reasonably tranquil conditions within the lagoon.

North breakwater is about 600 m long. It stands about 400 m north of the entrance, slightly inclined southward. South breakwater is a shade longer and stands about 400 m south of the entrance, slightly inclined northward. So between the two breakwaters, there is about 800 m of open coast. See the Figure 21.1 below.

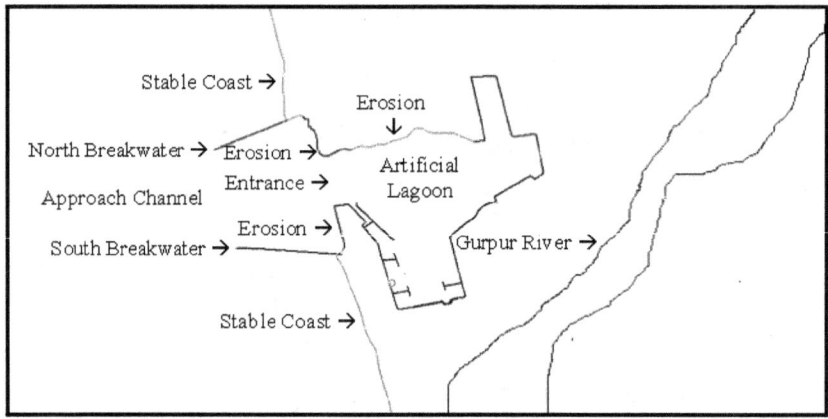

Figure 21.1: New Mangalore Port and Gurpur River

Experts felt the sediments that silted the port came from sea with the littoral currents, through the entrance. That is a generally accepted view. There are scientific papers about it too, by some eminent names in coastal and harbour engineering.

If the littoral currents were moving six million cubic metres of clay through the entrance bracketed by breakwaters, we can expect, at least, some clay to deposit either side of the entrance. Therefore, the two 400 m long stretches either side of the entrance must accrete or at least remain stable. But that is not the case. Both stretches are eroding. Seawalls have been built there in a bid to arrest the erosion, but to no avail. In addition, inside the lagoon, shoreline north of the entrance is also eroding. Even that should not happen if littoral currents were delivering sediments into the lagoon through the entrance.

A coast will erode only when the waves, even small ones, break on it, with no sediment supply getting through from land or from anywhere else. With adequate sediment supply no coast will erode. Therefore, the assumption that the littoral currents are delivering sediments into the lagoon is not tenable.

The approach channel is also silting profusely, every monsoon. Experts believe that longshore currents are delivering sediments into the channel. But on the Mangalore Coast, waves are head on, particularly during monsoon. Precondition for longshore drift is waves breaking at an angle to the coast. Therefore, there should be no longshore drift along the coast.

Say, there was such a drift along the coast, somehow, then, just like at Madras Coast after the pier, there should have been accretion upstream of one of the breakwaters and erosion downstream of the other, depending on the direction of drift. That too is not the case. Therefore, longshore drift is not the cause of siltation in the approach channel.

There was a glitch in the lagoon's excavation that probably went unnoticed. That was the thinning, if not breaching of the clay lining that separates the groundwater aquifers from sea. A breach in this lining normally does not cause seawater to seep into the aquifers, but the other way, because the aquifers are usually at a higher elevation. We can therefore expect the groundwater to seep into the lagoon.

Groundwater is not mud free. It could bring in some mud into the lagoon. But that may not be significant enough to account for the port's siltation. Therefore there must be a larger source of sediments. Excavation also put the Gurpur River flowing within three to seven hundred metres from the lagoon. As a result, the layer of clay between lagoon and river was thinned, perhaps even breached at some places. Therefore, water from the river could seep into the lagoon. During the non-monsoon months, little water flows through the river, so the seepage into the lagoon may be too small to make any difference.

As the monsoon rains start, the sediment-laden runoff flows into the river copiously from its basin. The water level in the river increases suddenly and sharply. Even its sediment load increases significantly. As the river's water level along the lagoon rises, the hydrostatic pressure acting on the already thinned clay lining also increases. That causes the lining to breach forming subterranean channels, through which the river literally drains into the lagoon, bringing with it huge amount of sediments, mainly clay. We can confirm that by a radioactive tracer study. Scatter the tracers on the riverbed, along the lagoon, at the start of monsoon. The tracers can soon be tracked within the lagoon. Further tracking will show the tracers flowing out through the entrance into the approach channel. That is how the approach channel silts.

The river's draining into the lagoon during the monsoon also can be confirmed by measuring the salinity within. Water in the lagoon would be significantly less brackish, almost fresh. But water samples should be taken few meters below the surface. On the surface, the tidal stream would keep the water saline. After the monsoon, as the water level in the river recedes, the subterranean channels seal themselves shut with the clay in them congealing. That brings to end the siltation. Water in the lagoon soon turns salty.

Can we prevent the port silting? When the problem is understood, solutions do emerge. The aim is to prevent the river draining into the lagoon. There may be many ways to achieve that. A workable solution can be hammered out by hydrologists, hydro-geologists and civil engineers putting their heads together.

CHAPTER 22

NOURISHING BEACHES

Up to three quarters of the world's beaches are eroding. General view, among the experts and common folk alike, is that waves are causing the erosion. That made concepts like Kadalakramanam or sea-attack popular. But the waves and coast have coexisted long before erosion became an issue. So what is causing erosion? The answer may be quite obvious already.

The waves breaking on the beach remove the fine sediments, unceasingly with every run-back. But that becomes a case of erosion only when the beach stops getting the sediment supply. Therefore, disrupting the sediment supply to the coast is the only cause for erosion. There is just no other reason, anywhere in the world. The sediment supply can be disrupted both naturally and by human intervention.

Land erosion ceaselessly creates sediments, which are delivered to the coast mainly by runoff, flowing either directly across or through the natural channels of rivers, streams and creeks. Sediment supply to the coast will therefore depend on the rainfall on the adjoining landscape and also on the pattern of runoff drainage towards the coast.

What if the region stops getting rain, say, due to a natural climate change? The result would be little runoff to deliver sediments to the coast. So the coast would erode. South Kathiawar Coast is one such naturally eroded coast. Long ago, this coast was almost entirely sandy, a hundred and fifty kilometres long beach! It began to erode when the Kathiawar Peninsula stopped receiving rains, after it went out of the scope of the monsoon. It then did not take long to turn rocky.

Sediment supply also can be disrupted when the runoff drainage is altered due to changes in the gradient of landform adjoining the coast. That can happen either gradually or suddenly. Gradual changes happen due to uneven land erosion. Earthquakes can cause sudden changes. Headlands, coastal cliffs and rocky shorelines on the Indian Coast resulted when the sediment supply was cut off due to changes in runoff drainage through uneven erosion of the adjoining landscape.

Humans became agents of coastal erosion only recently. But they are extremely efficient at that. Nowadays, they do so in more number of ways than ever before. Roads and rail embankments along the coast disrupt the sediment supply. Coastal townships also disrupt sediment supply. Dams and barrages bring down sediment supply to the coast. Jetties, piers, groynes, breakwaters and even entire ports impede the longshore drift. Result is rampant erosion, world over. In fact, every case of coastal erosion today is manmade. In other words, there is no case today of entirely nature induced coastal erosion!

Ominous aspect of coastal erosion, natural or manmade, is that once it sets in, it is almost unstoppable. The only way to stop erosion would be to restore the sediment supply. Once disrupted that may be extremely difficult to restore.

Without the sediment supply, the breaking waves will go on eroding the coast until all the sediments, both from the coast and the adjoining landscape, are removed to sea. The process would go on till the rocks are bared. Breaking waves only strip away the sediments. In the process, loose boulders may roll down to sea, but the large rock formations that run deep would remain standing. Every rocky or cliffy coast that we see today is the result of erosion. That is also the future of every coast that is eroding today!

If the large rock formation is deep inland, erosion will go on till waves meet up with the rocks, in the process create a bay. Erosion almost formed a bay north of Chennai Port. That is where the Chennai Fisheries Harbour has been built. One of the natural bays formed entirely by erosion is Palk Bay. I shall deal more on this later in the last chapter, the Chapter 47: Brief History of the Monsoon Coast.

Every nearshore island or islet is also the result of coastal erosion. After a rock patch along the coast is stripped bare, erosion sometimes continues landward from its sides. Soon the rock patch is separated and becomes an island or islet, depending on the size.

No matter how the sediment supply is disrupted, the erosion starts at the surf zone, where normally the waves break. That is also where the run-backs are swiftest. As fine sediments are removed, with no fresh supply getting through, the surf zone becomes steadily steep. Soon, even the coarser sediments begin to roll down to sea. Before long, the surf zone becomes a vertical wall of mud.

Waves then break directly on this wall, steadily scouring it, and pushing it landward. As it retreats landward, it gets higher. Waves then break at its base. As the base erodes, an overhang forms above. As the erosion progresses at its base, the overhang gets bigger and soon unsustainable. It therefore collapses making the wall vertical again. Erosion resumes at the base to create yet another overhang. The process goes till waves meet a large un-erodible rock formation inland. The process of erosion is as simple as that. Is there a solution?

Except in exceptional cases, it is extremely difficult to restore the natural sediment supply to the coast. That will warrant dismantling whatever that disrupted the sediment supply. In addition, the eroded coast will have to be restored to its original form and gradient. Theoretically that may be possible, but the cost and effort can be prohibitive. Therefore, at best, we can put in place a temporary solution, which then will have to be reapplied at regular intervals.

A modestly workable solution for restoring an eroded beach is known as beach nourishment. It involves delivering offshore sediments directly on to the beach. That is where all the eroded sediments have gone anyway. It requires a special dredger that can excavate the seabed and pump the sediments onto the beach, in the form of slurry, across a wide arc. It is an impressive sight. The resulting spray refracts sun's light into multitude of rainbows. So the experts fondly call it 'rainbow technique'.

The sediment concentration in the slurry pumped onshore is usually about 15 to 20%. Rest is water, which flows back to sea. Material pumped onshore may appear quite muddy, but that is not a problem. Run-backs would soon strip away clay and silt to render the beach sandy. The process would then need to be repeated as the beach erodes, unless the sediment supply is somehow restored, naturally or otherwise.

When a port impedes the longshore drift, the coast downstream almost immediately starts to erode. Experts believe that the erosion downstream can be managed by pumping sediments across the port through a floating pipeline. They call it sand bypassing.

The concept is based on the myth of longshore currents moving sediments along the coast. So they believe that a large offshore pit dredged upstream of the port would capture most of the sediments thus moved along the coast. It is the same logic applied to measure the longshore drift with a sand-trap. Recall the discussion in the Chapter 6: Mythical Currents.

Sediments from the pit are then pumped up and delivered downstream, across the port, through a floating pipeline. Actually, sediments that fill the pit are what run-backs deliver from the coast across the pit. That is a trifle compared to the demand downstream, that is, what the run-up driven longshore drift naturally transports along the coast had there been no impediment. Therefore, sand bypassing can never make good the loss due to erosion downstream.

The sediments lost from an eroding coast go to sea right across, and nowhere else. Therefore, instead of bypassing from the upstream, the sediments are best recovered offshore. That is beach nourishment anyway!

Sand bypassing has been installed at the Vishakhapatnam Port, on the East Coast. A massive trestle pump has been erected over a pit south of the port. The sediments from the pit are pumped up and delivered north of the port through a floating pipeline. But that does not go on throughout the day, because the pipeline has to be frequently severed to allow vessels in and out of the port. The arrangement has not worked. The coast downstream has eroded completely. A massive seawall now extends north almost up to the Ramakrishna Beach.

The experts involved however claim that the Ramakrishna Beach has not eroded, because of their sand bypassing arrangement. That is not quite tenable. The bypassed sediments just cannot reach the beach, because there is a natural barrier, a rock-outcrop, at the southern end of the beach that separates it from the eroding coast south. That was quite fortunate. Otherwise the beach too would have eroded. That also means the beach did not get sediment supply from the south, via the longshore drift, even before the Vishakhapatnam Port was built! How then did the beach get its sediment supply?

The answer lies in the local topography. The beach is fringed by a hill along its length. Whenever it rains on this hill, sediments flow down with the runoff to replenish the beach. The supply went down after the Beach Road was built, though not entirely, because road engineers had slotted in a number of drains, at regular intervals, below the road. Sediments now reach the beach through these drains, though not as copiously as before.

To test the hypothesis, we must look for an obstruction that may have come up recently, between the road and beach. There is in fact one right on the beach, the Kursura Submarine Museum, a decommissioned submarine hauled up on the beach and turned into a museum. The submarine rests on concrete chocks. A masonry boundary wall has been built around it. The wall cut off the sediment supply to the beach seaward. As a result, there is greater depletion of the beach right across the wall. The problem can easily be remedied by either making openings at the base of the wall to allow sediment flow or replace the wall by a chain-link fence supported on pillars. That will restore the sediment supply. The depleted portion would soon grow back to get in line with the rest.

CHAPTER 23

PONDICHERRY BEACH ROAD

In 1998, I was sent on temporary duty to Pondicherry. It was my first visit there. The purpose was to invite Professor Kittu Reddy, an acclaimed motivational speaker from Aurobindo Ashram, to deliver a series of talks at the Goa Naval Area. I was put up at one of the Ashram's guesthouses.

On the first day, around sunset, I went on a stroll along the Beach Road, the Goubert Salai. Along the road was a sandy stretch, which I presumed was a beach. After all, a beach next to a Beach Road is what one must expect. At the southern end of the road was a long pier. It reached into the sea, perpendicularly. I was reminded of the 1876 Madras Pier. Pondicherry Coast experiences similar wave climate, therefore north of the pier too must erode. Sure enough, the beach along the road had long eroded. In its place was a massive seawall. A layer of sand was spread over the seawall to camouflage its ugliness.

Interestingly, there was little accretion south of the pier. It is a lattice structure that did not contain the sediments that longshore drift deposited on its upstream. So the beach south stayed narrow. If it had been a rubble-mound or masonry structure, surely there would have been a wide beach there, like Chennai's Marina Beach, probably wider.

Now, even the narrow beach upstream of the pier must be eroding. There is another impediment to the longshore drift further south, the Pondicherry Harbour. The harbour is within the estuary of Ariyankuppam River, but there are two breakwaters on the coast, one on each side of the river mouth. As a result, the coast south has accreted. Erosion north is therefore a foregone conclusion.

The seawall along the Pondicherry Beach Road was securely built, unlike most other seawalls elsewhere. Building the seawall normally meant dumping boulders along the eroding stretch, usually from a tipper-truck. But is this securely-built seawall serving its purpose?

One of the readily apparent signs that a seawall is not protecting the adjoining landscape is subsidence or sinking of land along its length. Why that happens, I shall tell you in a moment. Sure enough, the Beach Road showed signs of subsidence. There were several cracks on it, which of course were patched up with bitumen or concrete. It is only a matter of time before the stately buildings along the road begin to show effects of subsidence with cracks on their walls and foundation. Perhaps, there are cracks already.

Besides subsidence, this seawall itself did not appear to be holding out. As a whole, it appeared to have sunk considerably. The boulders on its seaward edge too have fallen away, at many places. Why is a seawall unable to protect the coast and even itself?

Everyone assumes that a seawall made up of huge boulders would withstand the persistent onslaught of waves. Designs are often evolved and tested in wave flumes. Prototypes invariably withstood the most intense waves that they simulated. But even such tested seawall crumbled in wave conditions much less severe. It is only a matter of time! Boulders do not crack up. Each one stays intact. But the seawall as a whole crumbles!

Let us therefore take a closer look at a robust seawall, and see what happens when the waves come crashing on it. Do not get too close. It may be dangerous. Boulders may be slippery too. Besides, the barnacles may tear the soles of your feet, even through the shoes. Since it is not practicable to observe what may be happening, let us do a 'thought experiment', wherein we can get as close to the seawall as we want, even go into the crags between the rocks, and also behind it.

When a wave crashes on the seawall, the run-up surges almost vertically up. Immediately, the wave-water rushes down as run-back. It strikes the bed along the seawall's foot with considerable force. After hitting the bed, the water surges along, because of its momentum, before merging with the sea. By then the next run-back comes crashing down.

The waves breaking on the seawall, no matter how intense, cannot even shake the boulders. But the run-backs hitting the soft bed and surging along it, one after another, scour away the sediments from the seawall's foot, bit by bit. It is then only a matter of time before a long pit forms there, into which the boulders slip down, one by one, bringing the seawall down. It may take some time for a massive seawall to collapse, but surely it will. Nature is not in a hurry.

In addition, as the waves crash on the seawall, some water, literally, rifles through the crags, all the way to the soil behind the seawall. With wave after wave crashing, the amount of water rifling through the seawall can be substantial. That soon soaks up the soil. It does not take long. When the soil is fully saturated, water begins to flow back to sea, through the seawall and along its base. Even this flow back scours away sediments from behind and below the seawall, grain by grain. That leads to subsidence at the adjoining landscape, besides causing the seawall to sink as a whole.

Where there is longshore drift, the seawall does even more harm. Besides offering little protection to the coast, it hastens the progress of erosion downstream. It behaves like any other obstruction that impedes the longshore drift. So to counter the spreading erosion, the seawall would need to be extended downstream, steadily, until the next sediment source, a river mouth that discharges sediments.

There is yet another 'hard solution' applied to tackle erosion, the groyne. A seawall comes between waves and coast, whereas the groyne prevents waves crashing on the coast directly, at least, in full force. Like the seawall, even building a groyne is no hi-tech procedure. It is dumping of rocks from a tipper truck, but perpendicular to the coast or nearly so. A single groyne will not do. A number of them will be needed, in a pattern, with sandy stretches between each pair. That is the groyne-field.

No matter how close the groynes are, the coast between will experience at least some mild wave-action, due to the wave refractions. That may be sufficient to erode the coast, perhaps slowly, if no sediment supply is getting through. Therefore, even the groynes cannot stop erosion. They can at best slow it down, like the seawall.

Like the seawalls, even the groynes do collapse. Often, as the erosion progresses inshore, the groynes are detached from the shore, to become small islets. Eventually these islets sink or submerge.

In addition, wherever there is longshore drift, even the groynes will cause the advance of erosion further downstream. But unlike a seawall, groyne-field offers some respite to the fishermen. They can continue to launch their canoes or catamarans through the sandy stretches between the groynes, though not for long. Soon the erosion will force building of seawall between the groynes. You can see that north of Chennai Port.

Some coastal experts believe that the groynes help build back the eroding coast. That is a myth. The eroding coast can grow back only if the sediment supply exceeds the removal by run-backs. There is just no other way.

CHAPTER 24

KAKINADA DEEPWATER PORT

In September 2007, I went to Kakinada, a port town on the East Coast, on a 'site visit' in connection with a hydrographic survey. An industrial consortium was planning to build an 'offshore' port, about ten kilometres north of Kakinada, and about ten kilometres offshore, in deep waters. The port would be connected to land through an elevated transit system—road and rail on stilts, with conveyor belts and oil pipelines running along.

A hydraulic model was needed to evolve and test the port's design. The model can be built only from a large-scale navigational chart. But the project site fell outside the limits of the available large-scale navigational chart, the Approaches to Kakinada, Chart 3009. Hence the site could only be plotted on the coastal passage chart from Sacramento Shoal to Kalingapatnam, the Chart 354. Chart 354 was on scale 1:300,000, which was too small a scale for building the model. So the designers sought a fresh hydrographic survey of the site. That was how I got involved as a consultant.

Before setting out, I was shown the Chart 354 with the site marked, a tiny pencil-drawn rectangle. As we were driving to the site, I casually mentioned to one of the design team members sitting beside me that the coast there may be eroding. I had only suspected that.

Soon we were driving along the coast north of Kakinada Deepwater Port. The coast was eroding, but there was a seawall, nothing but haphazardly dumped concrete blocks that offered little protection to the coast. Moreover, it had collapsed at many places.

The member who was sitting beside me, pointing to the crumbling seawall, said, "You see Commander, wherever there is some erosion, measures have been put in place to control it. There is no cause to worry about our project site. No erosion has been reported there."

I did not comment. I was yet to see the site. Obviously, it was their first visit too. Though the drive was along the coast, because of the intervening vegetation, we could not see the coast. On reaching the site, we left the car on the road, and walked a short distance across a sandy stretch to the coast. There, we came in for a surprise, perhaps shock.

The coast had already eroded, enormously. Instead of a coast with gentle gradient, we were standing on a steep cliff of sand. We could not get off the cliff edge, nor go too close. Sandy edge would have collapsed, taking us several meters down to the waterfront. So we stood at a safe distance and watched in awe. The waterfront below was strewn with fallen trees, mostly coconut palms. There was no seawall. From where we stood, the eroding coast stretched both north and south, as far as the eye could see.

The design team had a hand-held GPS. It was turned on to get a 'fix' of the coast. Now, they were certainly shocked, and so was I. From the charted coastline, we were standing more than two kilometres inland! In other words, over two kilometres wide swath of the coast stretching kilometres either way had eroded.

No one had paid a visit to the site before giving a go ahead for a port project that probably would cost billions. If a large-scale chart was available, no matter how ancient, even this site visit would not have happened. The designers would not have sought a fresh hydrographic survey to build their model.

The hydrographic survey of the site was completed and the data was handed over to the designers. What may have become of the 'offshore' port project, I cannot tell. But what caused the erosion? Obviously, something impeded the longshore drift, somewhere south.

Kakinada Deepwater Port is on the western shore of Kakinada Bay, in the lee of Godavari Sand Spit. See the Figure 24.1 below. As such, there are no waves within the bay during the Southwest Monsoon. Without waves breaking on the coast, there can be no longshore drift. Therefore, the port by itself could not have caused the erosion along the coast north of the bay. But something did cut off the sediment supply to the coast. What was that?

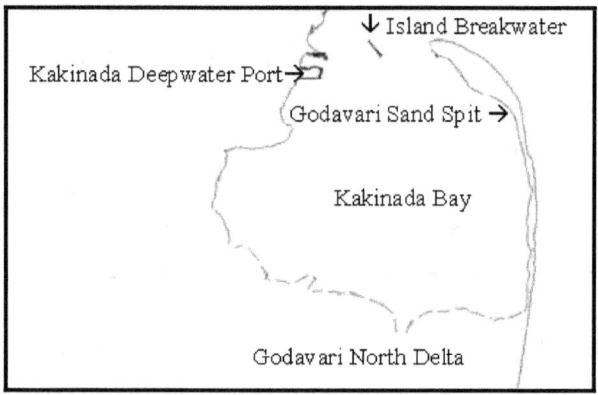

Figure 24.1: Kakinada Deepwater Port

To get to the answer, we must trace the origin of Kakinada Deepwater Port. Until 1970s, Kakinada was only an 'all-weather' lighterage port. It was so since the days of British. Ships came to anchor northeast of the bay, beyond the Godavari Sand Spit. Lighters or small barges transhipped the cargo between ships at anchor and port. The process was slow and cumbersome. But that was how the port operated since long. There must have been a good reason for that.

The River Godavari drains through a large delta. This delta is in three parts—northern, eastern and southern. Northern Delta drains directly into the Kakinada Bay. It brings in huge sediment load into the bay. That kept the bay perennially shallow. Therefore, only the lighters could operate, hence Kakinada stayed a lighterage port, but with an 'all-weather' tag, because of the shelter provided by the Godavari Sand Spit.

Eastern and Southern Deltas also deliver huge loads of sediments to the coast, which are then driven north along the Godavari Sand Spit by the longshore drift. The drift continues north across the shallow bay mouth to the coast north, along with the sediments discharged through the bay by the Northern Delta. Therefore, the sediment supply to the coast north was indeed huge. That was the situation until the 1970s.

Then someone had a seemingly brilliant idea. Why not bring in big ships into the bay and berth them alongside? The ships can then be loaded or unloaded quickly and cheaply. The lumbering lighters can be dispensed with. Kakinada could become a major port. CWPRS was tasked to design the Kakinada Deepwater Port.

Creating the port, with its operating basin, turning circles, wharfs and cargo handling facilities along the bay's western shore was simple. But the facilities would serve no purpose, unless the ships could enter and berth. The bay was too shallow for that. A dredged channel from the port to the deep waters beyond Godavari Sand Spit thus became necessary. See the Figure 24.2 below. The channel destroyed the coast north of the bay!

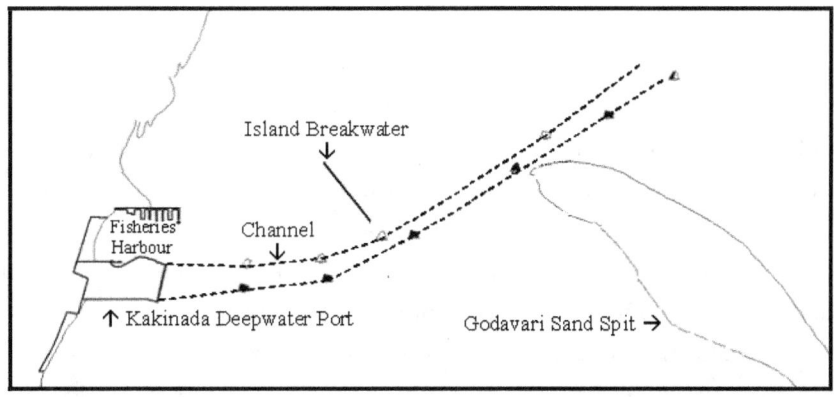

Figure 24.2: Approach Channel of Kakinada Deepwater Port

Sediments discharged through the bay and what came across the spit via the longshore drift went into the channel. Nothing went past to the coast north. The channel became a sediment trap. It would have silted to the brim in the first Southwest Monsoon that followed capital dredging, and in matter of few days. That would have restored sediment supply to the north coast. But that was not to be. The channel was continually maintenance dredged thereafter. For the port to remain open, the channel needs dredging, practically year round.

Therefore, it does not have to be a structure that stands proud of the seabed, like a breakwater or pier, to impede the longshore drift. A dredged channel will do just as well. Breakwaters, piers, seawalls and groynes are however the main disrupters of longshore drift. These features abound on the East Coast, from Nagapattinam Port in the south to Paradip Port in the north. The coast north of every such obstruction is eroding, with no end in sight. It is in fact unstoppable. So you can imagine how the East Coast would appear few decades later. Notwithstanding, many more ports have been planned on this coast. Construction of some is already underway.

If the ports on the East Coast are asked to pay for the damage they have caused to the coast or to restore the eroded stretches, most will go bankrupt! Who then will pay the price, the traditional coast dwellers, who are displaced when the coast erodes, or the taxpayers, who are either ignorant or unmindful of whatever is happening to the coast?

It is time to draw up a national policy to protect the coast, than to merely create avenues for coastal and port development through specially enacted instruments. Unlike the economic developments inland, it is not only about the people affected and their lands being acquired, but the very survival of the coast and the adjoining landscape.

For example, the situation north of the old Gopalpur Port on the Odisha Coast is so very precarious that the erosion of the adjoining coastal dunes, which has already depleted considerably, would send seawater flooding deep inland for tens of kilometres in the very next cyclone that passes that way. It is not something that may happen many years later, but perhaps very soon. And then it is only a matter of time before the cultivable land falls prey to erosion.

CHAPTER 25

DWARAKA TO SOMNATH

The most prominent 'natural' feature on the 300 km long West Kathiawar Coast is the rocky headland, the Dwaraka Head. It is about 30 km south of Humani Point, the northern end of the coast. Adjoining the headland is a tableland, where stands the town of Dwaraka, one of the important centres of pilgrimage for Hindus.

Marine archaeologists from Archaeological Survey of India found remnants of an ancient settlement, entirely underwater, off Dwaraka Head, which many believe that to be the City of Dwaraka founded by Sri Krishna, who is revered as the Avatar of Vishnu. He was also the central figure in the epic of Mahabharata. The epic, towards the end, vividly describes how the sea overran the city, suddenly and catastrophically. There are now several theories about how Dwaraka went underwater.

Some suggest that a sudden and catastrophic rise in the sea level sent the city under water. But there is no evidence of such a sea level rise anywhere else along the coast, enough to submerge an entire city. The sea level cannot rise only around Dwaraka. It must happen elsewhere too, in fact, all along the Indian Coast, and beyond.

Another theory suggests that a catastrophic storm surge during a cyclone destroyed the city. If that was so, the city's destruction was a real possibility, but not its submergence. After the cyclone passes, storm surge would recede and sea level would return to normal.

Some others believe that a tsunami may have destroyed the city. That too may be possible, but does not quite justify subsidence, unless the earthquake that caused the tsunami was right next to the coast and a massive one too. Such an earthquake could cause subsidence, but would leave tell-tale signs all around, as fault-lines and crustal deformities. There are none in the region to connect to a massive earthquake.

Moreover, during the cyclonic storm surge or tsunami, the city's destruction would have been sudden and chaotic. That would have left behind a trail of destruction. There would have been a huge find of everyday household artefacts, and also some human and animal remains in some fossilised form. The underwater ruins however appear to be in an orderly condition, with few such finds. Therefore, cyclone or tsunami being cause of the city's destruction and subsidence is not quite tenable.

It was, most probably, a simple case of coastal erosion. But the coast can erode only when there is a shortfall in sediment supply. There is no other way. Most probably, the city itself may have been instrumental in cutting off the sediment supply to the coast. It may have been a fort wall built landward of the city that cut off the supply.

Soon after the sediment supply was severed, erosion set in, but rather slowly. The city therefore went down slowly to sea, giving enough time for the inhabitants to salvage their belongings and move to safer places. They left behind only things they could not take away—remnants of few fallen walls, without any roofing, door posts or hinges. Even the bricks from their fallen homes may have been recovered for building their new dwellings elsewhere. So only the foundations were left behind. Erosion went on till the waves reached the rocks, which became a rocky headland, the Dwaraka Head. It stopped there. Dwaraka thereafter remained safe on the tableland above.

The coast south of Dwaraka is sandy, a beach, but with a steep gradient along the waterfront. That points to inadequate sediment supply. Adjoining landscape gets scant rainfall, only intermittently. Interval between sediment supplies therefore is long, during which waves scour away the waterfront rendering it steep.

All along the coast, slightly offshore, are several dead coral reefs. And beyond the reefs are small seabed mounds. These tell an interesting story. Kathiawar Peninsula did receive heavy rains long ago and for millions of years. The coast those days extended far more into sea. Along that ancient coast, landward, were large coastal lagoons or backwaters, similar to those along the Kerala Coast today. The peninsula then stopped getting rainfall, at least, as copiously as before. Why? I shall tell you in the Chapter 47: Brief History of the Monsoon Coast.

Reduction in rainfall resulted in reduced sediment supply to the coast. The coast began to erode. To start with, the erosion led to the fragmentation of the coast along the lagoons into islands, the barrier islands. Water in the lagoons turned completely salty, but stayed quite tranquil because of the shelter provided by the barrier islands. Corals began to grow in the lee of the islands. Corals can thrive only in sheltered seas, because that is where they get their nutrition, the marine algae.

Barrier islands are transient features. They have no provision for sediment supply from anywhere. Therefore, did not take long to erode away and become seabed mounds. With the shelter gone, the marine algae disappeared. Without nutrition, the corals too died, leaving behind the reefs.

The existing coast was the landward edge of the coastal lagoons that opened out to sea. It remained more or less stable because of the occasional sediment supply. But in the recent times, it has begun to erode. That demands an investigation.

The winds on this coast now are predominantly westerlies, the winds blowing from the west. The lay of the coast is northwest-southeast, therefore waves generated by the westerlies meet the coast at an angle. That drives the longshore drift southward.

We have already noted that wherever the longshore drift is impeded, while the coast upstream accretes, the downstream erodes. But that does not seem to be the case on this coast. While there is erosion downstream, there is little accretion upstream. That may be due to the steep gradient along the waterfront. The sediments that the longshore drift deposits upstream of the obstruction roll down to sea down the steep gradient, hence no accretion.

As we head south from Dwaraka, the first impediment is the massive breakwater of Porbandar Port. This port, not so long ago, was only a minor port and fishing harbour within the shallow Porbandar Creek. It was turned into an all-weather port after the breakwater was built. As a result, the coast downstream is eroding. Erosion has already brought the sea lot closer. The Porbandar Lighthouse, an imposing 41 m high round concrete tower, painted white with black bands, now stands barely 150 m from the waterfront.

About 50 km south of Porbandar is the Mangrol Fishing Harbour. This too was until recently a seasonal harbour. It became 'all-weather' only after the breakwaters were built. The result is erosion south.

Another 40 km south of Mangrol stands the Veraval Port. It is one of the important ports of Gujarat. This port too became all-weather only because of the breakwaters. Consequently, the coast south is eroding. This erosion may prove costly. The historic Somnath Temple is about three kilometres southeast of the port. Erosion has already brought the sea right up to the temple. It is the seventh built at that site. The earlier ones were brought down by humans. If the sea is allowed to take the temple, it may be for the last time. Another cannot be built at the same site. A seawall has since been built around the temple in a bid to protect it. That may be only a temporary measure.

CHAPTER 26

ROAD TO CHELLANAM

Let us return to Chellanam. The coast began to erode in the early 1960s. Experts held the littoral currents responsible. Local folk who witnessed waves wrecking their homes needed no theories. For them it was Kadalakramanam, plain and simple. Notwithstanding the theory or public opinion, a coast where the waves break can erode only when there is a shortfall in sediment supply. There can be no other reason. Therefore, something impeded the natural sediment supply to the coast, around the early 1960s. The timing is important!

Before we find what impeded sediment supply, we must establish how the coast got the supply before the erosion had set in. For a coast to remain stable it must receive adequate sediment supply, to make good the sediments being steadily removed by run-backs.

Monsoon waves are 'head on' to the coast. Therefore, the question of longshore drift does not arise. We can therefore infer that sediments came from across the coast. Landward of the narrow strip of land along the coast is the massive backwater, the Vembanad. This strip of land cannot be the sediment source. If it was so, it would have disappeared long ago, opening Vembanad to sea.

The sediments therefore must come from Vembanad alone. But the sediments being dense would remain on its bed. How can the sediments on the bed reach the coast?

When Vembanad fills up, as it does every monsoon, the excess water flows out to sea, through several shallow streams across Chellanam and elsewhere, besides what flows out through the principal outlet, Kochi Port. This floodwater flowing across Chellanam is sediment free! That means, for the coast to get its sediment supply, the sediments from Vembanad's bed must seep through the intervening soil. The soil at Chellanam is sandy, therefore can allow the seepage of sediment-laden water. But that would require some extra push, which must necessarily be in the form of hydrostatic pressure acting on the Vembanad's bed. And that happens when the Vembanad fills up during the monsoon.

As it fills up, the excess sediment-free water flows out to sea through the surface streams, whereas the sediment-laden water from its bed seeps through the soil to the coast. That is also when the monsoon waves lash the coast. The sediments that run-backs remove are quickly replaced by those seeping out—a quid pro quo. That was how the Chellanam Coast stayed stable until early 1960s.

Therefore, the impediment, whatever it was, came up between Vembanad and coast at that time. You may have guessed it. Yes, it was a road. But I was not quite sure after having heard so many esoteric theories about the erosion. I had to make a visit to Chellanam to lift the blinkers of preconceptions that I had acquired over the years.

The opportunity came only in 2001. I had to visit Kochi on temporary duty. After finishing the work I came for, I borrowed a jeep from INS Jumna, a surveying ship then based at Kochi, to visit Chellanam. It took about an hour through Kochi traffic to reach the Kochi-Chellanam Road. It was a good macadam road, nearly straight, with several 'humpy' culverts.

To the east was Vembanad. It appeared shallow, silted, like a wetland. Grasses were growing at many places. There was some cultivation too. At places, large patches had been reclaimed. Today, several houses and shops have been built along the road, after reclamation.

To the west of the road was a narrow strip of land, with few houses and vegetation. I could only hear the sea, but not see it. But I could see a seawall, intermittently, through the gaps between houses and vegetation.

There was no seawall across the culverts, to allow the streams flowing beneath to drain to sea. The local fishermen, the few remaining at Chellanam, launched their canoes from these stream mouths. Long ago, they had an entire beach to themselves, to park their canoes, to sort their catch, to mend their nets, and above all to regale in their lively coastal folklores and ballads.

In 2001, the road ended at Chellanam. But today it goes north up to Fort Kochi and south up to Alapuzha, mostly running along the coast. On that day, I stopped where the road ended for a 'glass' of tea at the 'chaya-peediga', the local tea shop. I struck a casual conversation with some locals. Some of them had lost their homes to Kadalakramanam. I did not tell them that the road was responsible for the erosion. They would not have believed me. For them, it was Kadalakramanam, because they had witnessed the sea attacking and tearing down their homes.

The soil at Chellanam is sandy. During monsoon, it gets wet and soggy. To go to anywhere, one had to trudge through the soggy sand. It was almost impossible to drive on. People had to walk long distances to board a bus. For all practical purposes, during monsoon, Chellanam was cut off from rest of Kochi. After persistent demands by the residents of Chellanam, the Kerala State Public Works Department built the Kochi-Chellanam Road in the early 1960s. It was not a simple matter to build the road on sand that stayed soggy during the rains. It required a secure foundation, deeper and firmer than elsewhere.

The road engineers built culverts over the streams that flowed out of Vembanad during monsoon. But they were not aware of the sediment-laden waters seeping out through the soil to the coast. The road's foundation arrested that. Vembanad began to silt along the road. Coast across starved of sediments began to erode. Erosion was dramatic and immediately apparent, whereas the siltation took time to manifest. Silted areas along the road were slowly reclaimed for cultivation and housing. When the coast began to erode, no one thought of the road. After all, it had improved their quality of life. Instead, waves got the blame—Kadalakramanam!

Later that evening, I visited a friend at Ernakulam, a handicrafts dealer. I mentioned to him about my visit to Chellanam. I was surprised to learn he too was originally from Chellanam. Interestingly, shortly after the Kochi-Chellanam Road was built, his father bought a car, a second hand Morris Minor. Two years later, they lost their home to Kadalakramanam. With the compensation they received, they moved to Ernankulam.

Before the erosion took its toll, the stretch of land at Chellanam, between Vembanad and sea, was about two to three kilometres wide. Now, at many places, it is barely five hundred metres wide. What if it breaches somewhere along this stretch? The possibility cannot be ruled out.

A massive flood in Vembanad could do that. Such a flood, though rare, cannot be ruled out. The last major flood in Vembanad took place in AD 1341. What happened during that flood? That is another story. See the Chapter 28: Muziris to Kochi.

Only few short stretches on the West Coast receive sediment supply from the adjoining backwaters, pushed out by the hydrostatic pressure, because there are only few remaining. Rest have silted up into land. The coast either side of the river mouths get the supply from the rivers after deflection by sandbars. Even that accounts for only a small length of the coast. What about the rest?

The landscape adjoining the West Coast receives heavy rains during the monsoon. The sediment-laden runoff flowed to sea through numerous small rivers and streams, and also as overland flows across the stretches, where there are no rivers or streams. The overland flows however pass unseen to the coast, seeping through the sandy soil. Unlike the seepage out of Vembanad, here there is no need for any hydrostatic pressure to push the sediment-laden water through the soil. The natural seaward slope of the landscape provides the necessary motive power. That was the situation until the coastal roads were built.

When constructing the road, the engineers built again culverts over the shallow streams. Largely sediment-free water flowed out to sea beneath the culverts. The sediments that normally seeped through the soil however remained trapped landward of the road. As a result, the coast starved of sediments begins to erode as soon as the monsoon sets in and the waves begin to lash the coast.

When the coast begins to erode, solution in every case is the seawall. But no one takes notice of the fact that where there is no coastal road, at least, near the coast, there is little erosion.

Today coastal roads fringe most of the Malabar Coast—Kerala, Karnataka and Goa. As a result, the coast erodes, every year, as soon as the monsoon sets in. In some cases, the sea has reached the road. There are however few coastal roads on the Konkan Coast south of Mumbai. The terrain along the coast is not quite conducive to long coastal roads, because of the cliffy headlands jutting into sea, with only short beaches in between. But north of Mumbai, the coastal landscape is again flat over long stretches. So, the coastal roads have been built there too. As a result, the coast is eroding, though mildly, because the monsoon waves are not as intense as on the Malabar Coast.

Coastal roads are cheaper to build, because of the level terrain, and fewer obstructions. Moreover, everyone enjoys the drive along the coast. But the roads are destroying the coast. If the coastal road is absolutely essential, it must be built on stilts or laid over concrete pipes placed one beside the other.

Unlike on the West Coast, the coastal roads along the East Coast pose no problem, because little sediment-laden runoff flows across the coast, except in few places, where the hills fringe the coast, for example, the Beach Road along the Ramakrishna Beach, at Vishakhapatnam.

CHAPTER 27

DAM-N THE COAST

For past some years, the north bank of Narmada Estuary has been eroding, mildly though. Two things are clear from the erosion, the waves are reaching the eroding bank and there is a shortfall in the sediment supply. Both these must come together for the bank to erode. The estuary gets sediment supply from the river. Therefore, if there was sediment shortfall at the north bank, it must be so at the south bank too. But south bank is not eroding. That means waves are not getting to the south bank. A massive mudflat in fact shelters the south bank. Obviously, this mudflat does not get sediment supply from the river. It is sustained by the sediments that the Gulf of Khambhat discharges.

The sediment shortfall at the estuary usually happens when a dam is built over the river, more so, if the dam is close to the estuary. The nearest dam upstream of Narmada Estuary is the Sardar Sarovar Dam. It is about 125 km upstream, as the crow flies, or about 200 km, if you go along the river. That may not seem close, but here the distance does not matter. Very little water is released from the dam to downstream. Besides, there is little inflow into the river downstream of the dam. Without enough water flowing through the river, the sediments cannot be moved downstream. That leads to sediment shortfall at the estuary. The waves are breaking on the north bank, so it is eroding.

There are no easy solutions, because it is almost impossible to restore the river's discharge. Dumping boulders or concrete blocks along the eroding bank too is not a long term solution, but then there is no other in sight. The erosion will go on up to where the waves can reach, in other words, interminable. Fortunately, the process is slow. That is the situation when the dam is deep inland. What if the dam is right on the coast? There is one under serious consideration—Kalpasar Dam.

While we are still on dams and their impact on coast, I must tell you another story of a dam that shut down a functional port, literally, overnight, though not due to erosion. Until mid 1970s, Port Khambhat, situated at the head of Gulf of Khambhat, was functional. The port had been in existence since several centuries. It had extensive trading links with Middle East and East Africa. The main exports were salt, dried fish, ornamental stone beads and agricultural produce, which included cotton. Erstwhile rulers of Khambhat and surrounding provinces sailed regularly from the port on Hajj pilgrimages in their stately dhows. During the British era it was a designated as a minor port. The status continued even after independence. A rail line connected the port to Anand, one of the important centres of commerce in Gujarat.

Then suddenly, around mid 1970s, the port silted up. So it had to be abandoned. Nobody knew why the siltation occurred. Everyone assumed that it was, though mysterious, a natural phenomenon, with no apparent human intervention.

The siltation was unprecedented. A five kilometres wide swath along the port silted, with hard, sticky clay and gritty sand. That rendered the port's approach so shallow that not even small boats could operate. This once famous port is today left with only one motor boat. Even that remains at anchor about three kilometres away from the dilapidated Port Office. And from there, it can only be deployed during the high water springs, that is, once every fortnight. The port's rail line has gone. Tracks and sleepers have been stripped off. Only the white stones on the embankment remain to tell the tale.

The port is situated on the north bank of the estuary of Mahisagar River that drains into the Gulf of Khambhat, close to its head. The river is about 500 km long. It rises in the hills of Madhya Pradesh, and then flows through parts of Rajasthan, before entering Gujarat and flowing into the gulf. It is one of the three major rivers in peninsular India that flows from east to west, the other two being Narmada and Tapi.

Nothing of consequence happened anywhere in the estuary or at the head of gulf to explain the siltation. Therefore, something must have happened upstream. Whatever that was, must coincide with the onset of siltation. That was not difficult to find. The port silted up as soon as Kadana Dam was commissioned, most probably, in the first monsoon that ensued. But the dam is nearly 150 km upstream, as the crow flies, and lot farther if you go along the river. How did it contribute to the siltation?

Before the dam, the river discharged copiously every monsoon. The peak monsoonal discharge dictated the river's cross-section, width and depth, right up to the estuary. After the dam, the discharge reduced considerably and also became intermittent, that is, only when the dam released water. The reduced discharge flowed through a narrow channel on the original riverbed. Like Narmada, there was little inflow downstream. The estuary, at the gulf's head, was originally wide and deep, but that became unnecessary after the dam. As a result, both banks silted up to reduce that cross-section. But how did so much clay get there?

The dam not only reduced the amount of water flowing downstream, but also significantly altered the composition of its sediment load. Before the dam, the river discharged sand-rich mud. That was not sticky, hence flowed easily along the bed as the river journeyed to the gulf and beyond. After the dam, heavier sand largely remained within. Only the clay-rich mud flowed out whenever the dam released water, which happened only during the monsoon. Clay-rich mud is sticky. As long as the water flowed swiftly downstream, because of the steep bed gradient, the sticky mud kept flowing along. But as the river levelled out, at the estuary, the rate of flow reduced. The water laden with sticky clay-rich mud fanned out towards the banks. The mud stayed stuck to the banks thus reducing the estuary's cross-section, adjusting to the reduced discharge. Port Khambhat, on the north bank, therefore had to be shut down.

Gujarat Maritime Board has serious plans to restore the port to its former glory. Dredging the clay may not be easy, but not impossible either. The siltation process has since stabilised, hence, as it stands, no further siltation can occur. But soon after dredging, the port will silt again, as soon as the dam releases water. It is a no-go situation. Port Khambhat is finished! In any case, restoring the port will serve no purpose, because the gulf is all set to be dammed.

With that we can move on to yet another ancient port that may be heading for a closure, though not due to a dam.

CHAPTER 28

MUZIRIS TO KOCHI

Kochi Port is the oldest and still functional port on the Indian Coast. It came into existence shortly after the year 1341. That was the year when Vembanad flooded. Before the flood, and since the dawn of civilizations, the only port in the region was Port Muziris. It was within the deep and sheltered estuary of Periyar River, about 25 km north of the present day Kochi Port. But when Port Muziris functioned, there was no Port Kochi!

Muziris was famous in the then known world for the exotic spices—pepper, cardamom, cinnamon, nutmeg, cloves and many others. The port was connected to large parts of the spice growing Kerala through a network of navigable rivers, canals and backwaters—south up to Thiruvanthapuram, north up to Ponnani and east up to where rivers Periyar and Bharathapuzha were navigable. Spices came to Muziris by boats. The spice trade was brisk.

Sailing ships came to the port from the Middle East, running with the swift monsoon winds. 'Running with wind' means sailing with wind astern, literally, pushed by the wind. Muziris offered safe harbour from the rough monsoon sea. The sandbar across the mouth ensured tranquillity within the estuary. Local pilots knew how to navigate the ships in and out, across the sandbar.

One of the famous visitors to Muziris was St Thomas, one among the twelve disciples of Jesus Christ. He landed in AD 52 bringing Christianity to India centuries before the European missionaries.

Those days Periyar Estuary was the principal outlet of Vembanad, through a deep channel east of Vypin, which today is an island north of Kochi Port. Vypin then was not an island, but connected to the present day Fort Kochi. Probably, during the monsoon, a shallow stream flowed between Vypin and Fort Kochi, like those flowing across Chellanam. During the dry season, one could walk across from Vypin to Fort Kochi. In other words, there was no opening of Kochi Port. That was the situation until 1341.

In that year, during the monsoon, Vembanad flooded. The flood took many lives and altered the local geography. The surging floodwaters breached the coast between Vypin and Fort Kochi. Vypin became an island. That marked the end of Muziris, literally, overnight. The port soon went into oblivion. Spice trade was hit, but not for long. It soon resumed through the newly breached opening at Kochi, the new Kochi Port. But why did the flood shutdown Muziris?

After the flood, Vembanad began to drain through Kochi. As a result, little water flowed through the Vypin Channel, thence through the Periyar Estuary. That led to silting of both channel and estuary. The two had stayed deep until then only to drain the copious discharge from Vembanad. When that ended, there was little need for the two to remain deep. Siltation set in immediately to reduce the cross-section, adjusting it naturally to the reduced drainage. So, it was siltation that did Port Muziris in. Everyone soon forgot that such a port ever existed.

Muziris later came to be called Kodungallur. Europeans however called it Cranganore. The place stayed on in memory, because of its famous visitor, St Thomas. Centuries later, Portuguese made attempts to revive the port. They did not succeed.

Apparently, there are once again plans to develop a port within the Periyar Estuary. As long as Vembanad drains through Kochi, it will be impossible to keep such a port naturally deep. It will require extensive capital dredging to start with, and then regular maintenance dredging thereafter. Dredging would increase the drainage through the Vypin

Channel, thence through the estuary. That would correspondingly reduce the drainage through Kochi. The inevitable consequence would be increased siltation at Kochi Port. Kochi, Vypin Channel and Periyar Estuary are parts of the same homogenous water body, the Vembanad. Wherever it is dredged, drainage there will increase, with corresponding reduction elsewhere. The reduction of drainage will soon translate as reduction in channel cross-section through siltation. Therefore, it would be prudent to leave the present equilibrium undisturbed, if Kochi Port has to remain functional.

Kochi Port stays deep only because of the huge volume of water draining through it, as was the situation at Muziris before the flood. Let us now take on the tacky question—what will happen to Kochi Port, if Vembanad, say, after yet another massive flood opens up somewhere south, perhaps at Chellanam? Chellanam is the most vulnerable zone.

That will suddenly and drastically reduce the drainage through Kochi. The immediate consequence will be siltation, which could set in so rapidly that the vessels alongside would be grounded. Dredgers too would be unable to operate. Therefore, the need to shore up Chellanam Coast assumes immense significance. Seawall is not the solution.

A single day's rain, like that fell over Mumbai on 27 July 2005, may be sufficient to flood Vembanad. Meteorologists say it was due to a phenomenon known as the monsoon vortex. They felt that a series of such vortices somehow formed rapidly over Mumbai leading to the exceptional downpour—944.2 mm of rain in just 24 hours.

Though rare, phenomenon is impossible to predict. There is no reason why it cannot develop over Vembanad, any time. Most probably, the 1341 flood was due to something similar. Are we prepared?

Even without such a catastrophe, some recent developments may be slowly pushing the Kochi Port to the brink of closure, most likely through bankruptcy. The port may become unviable, because of the huge dredging bill.

Before I go into that, let us see how the present Kochi Port was developed. That is the story of Bristow's Harbour.

CHAPTER 29

BRISTOW'S HARBOUR

Soon after Kochi Port came into existence after the 1341 flood, ships began to call regularly for spices. Initially, the ships came from the Middle East. Later on came the Europeans, rounding the continent of Africa—Portuguese, Dutch, French and British, in that order.

British eventually wrested control of the Kochi Kingdom. With that they gained control of the port. The port at that time operated from Mattancheri on the eastern shore of Fort Kochi, the Mattancheri Wharf. Today, only fishing trawlers are berthed there.

The harbour at Kochi was deep even for the large steel-hulled steamers of early twentieth century. But the vessels could not enter the harbour, because of the sandbar at the mouth. Only small vessels, with draught less than three metres, could ride over the sandbar, that too, only during the high tide. Big vessels stayed at anchor, off the harbour mouth. Boats transhipped the cargo to and fro. Transhipment was possible only when the sea was calm. Therefore, during the monsoon, the port was practically closed for large vessels, in spite of excellent harbour within.

Opening of the Suez Canal increased the number of ships calling on the

West Coast. Lord Wellington, the Governor of Madras Presidency, felt the need to develop Kochi Port. In 1920, he brought Robert Bristow, a leading harbour engineer to develop Kochi Port. Bristow was at that time looking after the maintenance of Suez Canal. He was appointed the Chief Engineer of Kochi Port Department.

Bristow felt Kochi could become one of the finest harbours in the world only if the big ships could enter it every season. Challenge therefore was to permanently rid the sandbar from the mouth. Dredging was not an option, because in the very next monsoon, mouth would silt and the sandbar would return. Dredging would then become an ongoing activity. That not only would be expensive, but also meant poor design.

The channel through Kochi, since the flood, was the principal outlet of Vembanad. Every monsoon, six major rivers and several smaller ones delivered into the backwater a huge volume of sediment-laden waters, which then flowed out to sea through Kochi. Outlet from Vembanad into Kochi was however narrow, barely a kilometre wide. During monsoon, because of the massive inflow, the sediment-laden waters, literally, surged out through this outlet. On the other hand, the channel through Kochi, those days, was wide, in fact, nearly three times wider, like a bulge, which spanned the shorelines of Ernakulam and Mattancheri. There was no island between the two, probably only some patches of mud, the mudflats. Immediately after the bulge was the narrow mouth on the coast.

The bulge served a purpose. It diffused Vembanad's discharge to a manageable rate. Sediment-laden waters thereafter flowed out to sea through the mouth at a gentler pace. That ensured stability of shorelines around the mouth. But that also led to the formation of sandbar across the mouth. The sandbar then deflected the sediment-laden waters to either side, north to Vypin Coast and south to Fort Kochi Beach. That was nature's way of ensuring the stability of the two stretches.

Bristow realised that the only way to prevent the sandbar would be to increase discharge through the mouth. That was easier said than done. If the discharge rate increased, the shorelines within and at around mouth would erode, widening both bulge and mouth. That is nature's way of restoring the former equilibrium. With that, the discharge through the mouth would return to the old rate and so the sandbar would also return.

So he decided to make firm the shorelines within Kochi right up to the mouth by building stone embankments. The Vypin shoreline, around the mouth, was exposed to waves, so he probably felt, needed extra protection. So, in addition to the embankment, he decided to slot in few groynes. Probably, the groynes may have been introduced only to further diffuse the discharge.

Only after the shoreline fortifications were in place could he proceed to the next stage—to increase discharge rate. The only way to do that would be to reduce channel cross-section. In other words, reduce the bulge between the two shorelines, Ernakulam and Mattancheri. That can be done by reducing width or depth or both. But depth is critical for a port, therefore must be increased, even at the cost of width. Therefore, narrowing the channel, while simultaneously increasing the depth was the way forward.

That was not a simple matter. Final discharge rate had to be just right. If it not fast enough, the sandbar would form and remain. If too fast, shorelines would erode in spite of the embankments. That was where the brilliance of Bristow as a harbour engineer came in. He decided to reclaim an island between Ernakulam and Mattancheri. His genius was in the precision with which he arrived at the shape and size of the island and width, depth and gradient of the channels either side of the island leading right up to the mouth and beyond. I wonder if the computations and drawings have been preserved. He may not have done all that in the head, or probably he may have. Geniuses sometimes work like that! Or was there a hydraulic model?

So the plan was to reclaim an island in the middle of the bulge, using the material dredged from the channels, but only after fortifying the shorelines with stone embankments. By the end of 1920, that is, within few months of his arrival at Kochi, he was ready with the plan. It was immediately approved.

Bristow immediately placed an order for a dredger on a yard in UK. In the meantime, he went ahead fortifying shorelines. The dredger was ready by end 1925. It was named Wellington, after the Governor. The dredger reached Kochi in May 1926. By then the Governor, Lord Wellington, who envisioned the development of Kochi Port, was on his way to Canada, as the Governor General. He returned few years later to India as the Viceroy.

The dredger's crew worked tirelessly, round the clock, every day, for the next two years, to dredge the channels and to reclaim the island. The island was named Wellington Island. See the Figure 29.1 below. On 26 May 1928, the first ship, SS Padma, a large steamship by the standards of those days, steamed into the new Port of Kochi.

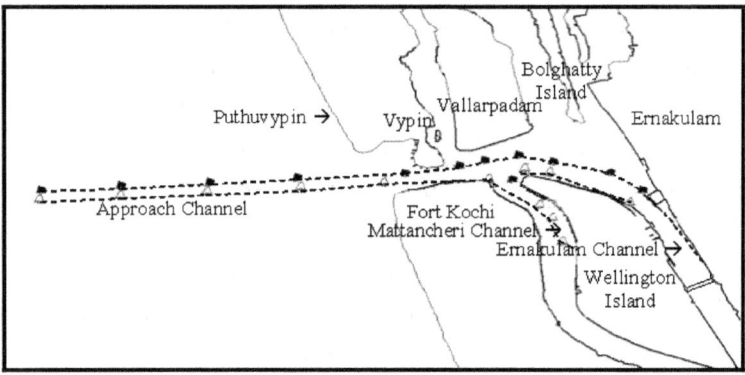

Figure 29.1: Kochi Port

As per Bristow's design, all shorelines, Ernakulam, Mattancheri and Wellington Island, were to be kept smooth. The embankments were therefore built with neatly flushed stones, with no projections whatsoever. Projections would impede smooth flow of sediment-laden waters. That would cause siltation around the projections, which could then spread into the channel. That also meant, every feature within the port that stood proud of the flushed-stone embankments were to be on smooth round piles or stilts, be it a small wooden boat jetty or a large one to berth ships, to ensure smooth, uninterrupted flow of sediment-laden waters through the channel.

In 1932, the Maritime Board of British India declared Kochi as a major port. The port was cleared for vessels up to a draught of thirty feet, that is, about ten metres.

For his remarkable achievement, Robert Bristow was knighted. In an address on the Radio BBC, on 11 Aug 1935, a proud and happy Sir Robert Bristow said—

"*I live on a large Island made from the bottom of the sea. It is called Wellington Island, after the present Viceroy of India. From the upper floor of my house, I look down on the finest harbour in the East.*"[6]

During World War II, Royal Navy took over the port. Wellington Island became a naval base, with its own airfield. Royal Navy built a jetty on the eastern shore of the island, the present day IN Jetty. They also built a boat pen, south of the jetty. As per Bristow's design guidelines, both jetty and boat pen stood on piles, to allow free flow of sediment-laden waters.

Royal Navy stayed in control of the port until the end of war. On 19 May 1945, the port returned to civilian control. The naval base continued to occupy a portion of the island. Rest went to the port. After independence, it became Indian Navy's second naval base on the western seaboard. Everything went fine with the port until the mid 1980s.

---

[6] http://en.wikipedia.org/wiki/Robert_Bristow

CHAPTER 30

FORT KOCHI BEACH DISAPPEARED

Right at the entrance of Kochi Port, to the south, was a beautiful beach, the Fort Kochi beach. Visitors, both locals and tourists, came to the beach in large numbers, everyday, to enjoy sunsets and to watch the ships sail past, in and out of the port. A portion of the beach ran along Navy's Gunnery School, INS Dronacharya, where I did my Sub-lieutenants' Gunnery Course in 1980. During the course, I spent some good time on this beach. Near the wardroom, there used to be a wicket-gate that led to the beach.

Up to the mid 1980s, the beach stayed stable, showing no signs of erosion. But today, there is no beach there. Instead you get to see an ugly seawall, with several groynes sticking out, like sore thumbs. INS Dronacharya is still there. The monsoon waves crashing on the seawall sent huge sprays of seawater on the buildings. The old wardroom has been destroyed. A new one has been built away from the seawall. Subsidence and water-logging too have been reported at the base. It may not be long before entire north Fort Kochi starts to get water-logged during monsoon. Perhaps, that may be happening already.

Why did the beach erode? Obviously, something impeded its sediment supply. But there was no construction of any kind on or near the beach to account for the disruption of supply. So, it became yet another case of Kadalakramanam. But that cannot be. There ought to be an obstruction that impeded sediment supply. We must find that. To get there, we must first know how the beach got its sediment supply when it remained stable. Erosion started around the mid 1980s. Timing is important.

Finding the sediment source is not difficult. The beach was at the port's entrance. Therefore, without doubt, the beach got its sediment supply through the port. Every monsoon, Vembanad discharged a continuous stream of sediments. The sediments flowed along the bed swiftly and unimpeded through the port. Sir Robert Bristow had ensured that by keeping channel sides smooth and obstruction free and the bed sloping adequately seaward.

The stream of sediments, soon after coming out of Vembanad, split into two sub-streams at the southern tip of Wellington Island. One flowed straight through the Ernakulam Channel and the other through the Mattancheri Channel.

Ernakulam Channel being in line with Vembanad's outlet, the stream that flowed through was swift and bore larger share of the sediment load issuing from Vembanad. The other flowing through Mattancheri Channel had comparatively smaller sediment load.

The streams bearing dense sediments flowed only along the bed. Surface flowing tidal stream did not affect them. Even when the tide was flooding, the streams kept flowing out, like the fluid-mud flowing out of Mumbai's mudflats. After passing the Wellington Island, both streams merged and flowed as one, unhindered towards the mouth. At the mouth, the swift flowing stream fanned out. The southern part of the fan went towards Fort Kochi Beach. That replenished the beach. The northern part went on to replenish the Vypin Coast. The middle part flowed straight out to the sea, swiftly, without forming the sandbar.

To make things clear, we can consider the stream flowing through Ernankulam Channel as made up of three separate streams, even though there are no discernible margins separating one from the other.

The two streams flowed along the shallow channel sides, one along Ernakulam shoreline and the other along the eastern shoreline of the

Wellington Island. Third one, which we shall call the 'Centre-stream', flowed along the deep mid-channel. Similar three streams also flow through the Mattancheri Channel.

The shallow stream that flowed along the Ernakulam shoreline turned sharply westward at Bolghatty Island, then flowed along the islands, Vallarpadam and Vypin, before turning north to the Vypin Coast. See the Figure 29.1 above.

At the Vypin Coast, the run-ups of breaking waves pushed the sediments on to the coast. That was how Vypin Coast got its sediment supply. The supply thus balanced the sediments being removed by the run-backs. The coast thus stayed stable. Since the stream replenished the Vypin Coast, let us call it the 'Vypin-stream'.

The other shallow stream flowed along the eastern shoreline of Wellington Island, unimpeded. Being swift it rushed past the mouth of Mattancheri Channel. As it did so, both the shallow streams issuing from the Mattancheri Channel joined it. The combined stream then flowed along the northern shore of Mattancheri to replenish the Fort Kochi Beach. So let us call it the 'Beach-stream'.

The mid-channel stream from Mattancheri Channel flowed into the Centre-stream flowing through the Ernakulam Channel, and together flowed swiftly out of the mouth without forming the sandbar. That was the precision with which Sir Robert Bristow had worked out the width, depth and gradient of channels!

So, if there is no Fort Kochi Beach today, something impeded the Beach-stream, somewhere upstream. What could be that impediment?

CHAPTER 31

SOUTH JETTY TROUBLES

Until early 1980s, the naval base at Kochi only had the IN Jetty that Royal Navy built during the World War II for berthing warships. As the Navy grew, more warships came to be based at Kochi. Berthing at IN Jetty soon became a problem. Extending the jetty was ruled out, probably because of its antiquity. So a new one south of IN Jetty was proposed. It came to be called the South Jetty. See the Figure 31.1 below. CWPRS designed the jetty after model studies on the tidal model of Kochi Port. The construction began in mid 1980s.

In December 1985, after my Long Hydrographic Course, I was posted to INS Mithun at Kochi. At that time, the South Jetty's construction was in progress. I wanted to take a closer look, only out of curiosity. But I could find time only at the end of the survey season, sometime in June 1986. There was little construction going on, because of the rains. I immediately realised that the jetty's design was rather odd.

At that time, I knew nothing about Sir Robert Bristow's design of the Kochi Port. That was not necessary. Every feature in the channel that stood proud of the shoreline was on stilts or piles, concrete or wooden—IN Jetty and Boat Pen on the Wellington Island and oil terminal and its mooring 'dolphins' on the Ernakulam shoreline, besides all the boat jetties.

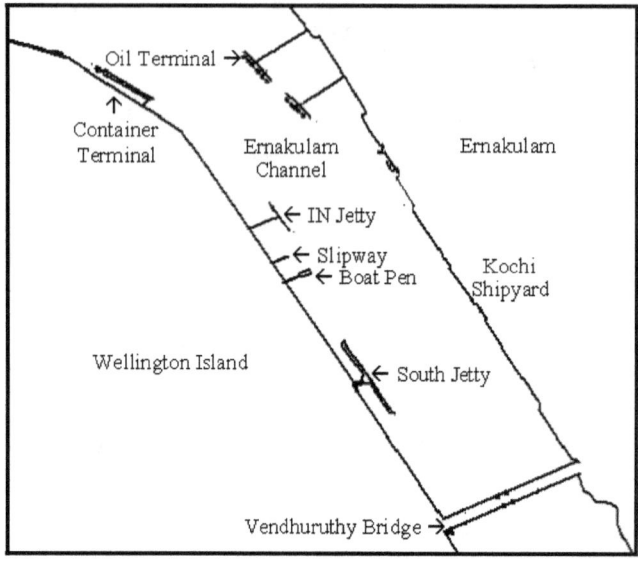

Figure 31.1: Ernakulam Channel

On the other hand, the South Jetty was a caisson-type structure, both its berthing face and approach from shore. The caisson is a large concrete box like structure that is built on a slip way or in a small drydock. It is then towed to the construction site, one after another, and weighted down on a specially prepared bed by filling it partially with rock debris and concrete.

The caisson-type jetty is easy to build and takes much less time compared to one that stood on piles, like the old IN Jetty, therefore lot cheaper. It is also far more durable and easy to maintain. Besides, the caissons can be used as tanks for storing fuel and freshwater. But in the Ernakulam Channel, it would be a disaster. Caissons would arrest the sediment flow. That would lead to siltation upstream. From the upstream, the sediments would slip into the berthing area. The siltation would then probably spread further downstream into the channel. That was what crossed my mind at that time. That did eventually happen!

Few days after the visit to jetty, I had to attend a 'monthly' meeting at the Headquarters, Southern Naval Command, in place of my Commanding officer, who was away on leave. At the end of the meeting, we were asked for any general points that were not included in the agenda.

I stood up, and said, "Sir, I think South Jetty's design is not quite right. It will cause serious siltation around the jetty and in the channel."

It was an audacious thing to do for a Lieutenant, who was only an officiating as a Commanding Officer. Everyone present looked at me quizzically.

Then one of the senior officers present spoke, "There surely will be some siltation arising out of any such marine construction. We have made provisions for maintenance dredging once the jetty is ready."

Most probably, he was associated with the jetty project, because he spoke with authority.

Chairman then asked me, "Are there any other issues with the jetty?"

I pondered a while before answering. All eyes were on me, but I took my time. Sediments flowing unimpeded along the Wellington Island eventually reached the Fort Kochi Beach, the next 'soft patch' downstream. All along, up to the beach, it was stone embankment. With jetty impeding the flow, the sediments would not reach the beach. I recalled what had happened after the Madras Pier was built. I had learnt it only recently, during the attachment to CWPRS, which was a part of the Long Hydrographic Course. It was still fresh in my mind. The coast downstream of the pier eroded, because of sediment starvation. The Fort Kochi Beach, also downstream of the South Jetty, would likewise face sediment starvation. Therefore, the beach would erode.

So I replied, rather confidently, "Sir, the jetty will cause the erosion of Fort Kochi Beach."

Everyone present probably felt I said something funny. A muted laughter ensued.

When the laughter subsided, the Chairman said, in not so lighter vain, "Young man, concern yourself with your ship. Leave the jetty to the experts."

The meeting was adjourned. I returned to my ship, taking care to avoid everyone. Thereafter, I got busy with ship's activities and hydrographic surveying. The jetty was not my concern anyway.

In 1989, I was transferred to Vishakhapatnam, and two years later to Mumbai, to the dockyard. There, one day, I ran into one of the officers, whom I had known since the Kochi days. He used to be one of the Staff Officers at the command headquarters. He was probably present at that monthly meeting.

He told me that my rather preposterous statement about the Fort Kochi Beach eroding was indeed right. The beach had disappeared. I had not known that until then. Whether the erosion was due to South Jetty or some other reason, he was not quite sure. No one, in fact, linked the erosion to the jetty. It was just yet another case of Kadalakramanam. There is more to the South Jetty than just erosion of the beach.

The jetty impeded the Beach-stream at the head of Ernakulam Channel. No stream, with or without sediments, thereafter flowed along the Wellington Island to the beach. As a result, sediments streams from the Mattancheri Channel slipped into the Centre-stream and flowed towards the mouth. That increased the sediment supply to the approach channel. The channel began to silt. A dredger now operates there continuously during the monsoon. If not dredged, probably, the sandbar would return, which Sir Robert Bristow had so assiduously prevented without dredging. In other words, South Jetty brought to naught all that Sir Robert Bristow had achieved!

Fort Kochi Beach got no sediment supply. That was obvious from its erosion and the eventual disappearance. The siltation in channel and upstream of South Jetty did not account for all the sediments that the Beach-stream bore before being impeded. Where did all the sediments go? In other words, what happened to the Beach-stream after it was impeded by South Jetty? There was a clue!

CHAPTER 32

PUTHUVYPIN-VALLARPADAM SEZ

As the Fort Kochi Beach eroded, the Vypin Coast accreted! But accretion was not quite dramatic as the erosion, that too, of a popular beach. It did not immediately attract public attention.

In 1988, one of the hydrographers, after coastlining the Vypin Island, reported that the Kochi Lighthouse appeared to have shifted inland. The lighthouse had not shifted position. The coast had accreted. The lighthouse, a 46 m high circular ribbed concrete tower, painted red and white bands, was built in 1979 close to the waterfront. Today, it is about half a kilometre inland!

A coast can accrete only when the sediment supply exceeds what the run-backs remove to sea. Therefore, the sediment supply to Vypin Coast had increased. It got the supply from Vypin-stream. Therefore, the sediment load of the stream had increased. But there was no commensurate increase in the amount of sediments discharged from Vembanad. How then did it happen?

After the South Jetty stopped the Beach-stream, some of its sediment load silted its upstream. The quantity involved was very small, because the available area upstream was small. Thereafter, the Beach-stream, still bearing almost its entire sediment load, was deflected to the Ernakulam Shoreline, which is only about 500 m away, like a river meandering away after it came up against an obstruction.

Thereafter, the Beach-stream flowed along with the Vypin-stream as one. Incidentally, both had similar flow characteristics, which made that possible. The combined stream therefore had nearly twice the sediment load. In other words, the sediment load of Vypin-stream, literally, doubled, thus doubling the sediment supply to the Vypin Coast. The coast began to accrete.

Two decades later, the accretion created a new landscape west of Vypin. It came to be called Puthuvypin or New Vypin. See the Figure 29.1 above. Puthuvypin turned out to be a bonanza for the land-strapped Kerala, a free-hold land, without titles or encumbrances. Everyone saw it as nature's gift. It soon became an attractive prospect for a port-based Special Economic Zone or SEZ.

The most difficult part of setting up an SEZ is land acquisition. Here it came free, without hassles. SEZ soon got underway. It came to be called the Puthuvypin-Vallarpadam SEZ, because it included the long since pending container transhipment terminal at Vallarpadam, east of Vypin.

Vallarpadam Island is an old landform, unlike the Puthuvypin. It remained unexploited until the SEZ was launched. Even Sir Robert Bristow did not think of exploiting it. Why? Simple, it lacked shelter from strong and often gusting monsoon winds.

Handling containers in wind is not only difficult, but dangerous too. Terminal would therefore have to suspend operations for many days during the monsoon. Vessels may have to be diverted. Resulting backlog could prove costly. In other parts of the world, the container terminals that lacked adequate wind shelter have reported problems of storage and backlogs, for example, Port Antwerp. But for those terminals, such wind related problems are only occasional. For Vallarpadam, it will be for up to four months every year, that is, for a third of the year. Will the terminal break even?

Even as the problems at Vallarpadam are unfolding, plans are afoot to create yet another shelter-less container port south of Kerala, the Vizhinjam Container Port. The project too has been on anvil since long. The proponents feel that it will challenge the hegemony of Colombo in container transhipment business. But the port's ability to handle containers during the four-month long monsoon and during the non-seasonal storms may be something they may not have considered. The project is however yet to take off, probably, due to funding constraints.

Puthuvypin came into existence because of an increase in sediment supply to Vypin Coast. What can happen if that is disrupted? What are the chances of that happening? Or has that already happened?

Vypin-stream originally flowed unimpeded along the shallow shoreline of Vallarpadam. The container terminal that has come up along this shallow shoreline is designed to handle large container carriers up to a draught of 12 m. The wharf therefore will require capital dredging at least up to 12 m or more, which must extend up to the main channel for the vessels to enter and leave the terminal.

After dredging the wharf, which has been completed, every monsoon, the Vypin-stream would discharge a significant portion of its sediment load, if not entirely, along the wharf. The wharf would therefore silt during the monsoon. From the wharf, the sediments will flow into the approach channel. That will further push up siltation in the channel, with possibility of the sandbar forming, unless continuously dredged through the monsoon. That in fact is the situation today!

The dredging bill may soon put the Kochi Port in red, if not already. Without continuous dredging, through the monsoon, the port cannot function. Sir Robert Bristow must surely be restless in his grave! Problems do not end there.

The sediment load of Vypin-stream further downstream will go down, significantly. In other words, the sediment supply to Puthuvypin will reduce. That would bring down the rate of accretion of Puthuvypin. But that may no longer be an issue. I doubt if anyone wants Puthuvypin to accrete anymore. But that is not the problem.

The depleted Vypin-stream may not even make it to the Puthuvypin Coast. Yet another obstruction has come up in the way, the International Bunkering Terminal. That will completely cut off the sediment movement further downstream. Hence, instead of accreting, Puthuvypin will erode. That too must have begun. What happens to the sediments stopped by the bunkering terminal? A wee bit would silt its upstream, but most would go into the approach channel, further pushing up the siltation.

The siltation in the channel can be managed by maintenance dredging, but what about the erosion of Puthuvypin? There are no easy solutions. But erosion may not be the only problem for Puthuvypin.

The monsoon winds traverse more than 8,000 km over the sea, drawing up huge amount of moisture from the sea all along the way. Along with moisture, winds also draw in huge amount of salt. On making landfall, the winds then deliver all that salt on to the coast. In other words, the winds literally spray salt on to the coast. Salt is extremely corrosive. It can eat into almost anything.

The traditional coastal folk know that. No wonder they build their dwellings only in the lee of dense vegetation or headlands. If you sail along the West Coast, you will be treated to a visual delight of dense greenery, but few buildings. On most other coasts around the world, people, particularly the rich and famous, live along the sea, in sea-side villas. Also, on those coasts are hotels, sea-side resorts, industries, power plants and more.

On the West Coast, only exposed structures were the lighthouses. There was no choice there. Mariners must be able to see them, un-obscured. The lighthouses on the West Coast therefore bear the brunt of salt-laden monsoon winds. The cost and effort that go into their maintenance must be higher than anywhere else.

Puthuvypin has no shelter from the salt-laden monsoon winds. Among the many things that have come up, the most worrisome are the steel tanks for storing petroleum products. It is a tank-farm, an array of several tanks. The salt-laden winds will therefore hit the tanks full blast for up to four months every year. The corrosion of tanks cannot be ruled out. Tanks will warrant elaborate maintenance. Corrosion will be severe during the monsoon. That is also when the maintenance most difficult. What if the maintenance, for some reason, does not keep up with the corrosion? The possibility cannot be ruled out, no matter how meticulous the procedures may be. Each tank, at any time, may be holding tonnes of one or other deadly liquid. What if one of the tanks cracked open and spewed its deadly contents, that too, during the thick of monsoon?

With that, let us return to Fort Kochi Beach, where the problem first manifest. The question is—is it possible to restore the beach and keep it stable thereafter? The answer is yes. In fact, it is a simple matter, unlike most other cases of coastal erosion, so much so, it may be treated as an exception.

All we need do is to restore the Beach-stream back on its former track. The beach will then grow back all by itself. We can hasten the process by some beach nourishment. Even for that there is no dearth of sediments. There is enough dredging going on already in the channel. Instead of dumping the material at sea, the dredger can pump sediments onto the beach. Even dismantling the existing seawall may not be necessary, but the groynes must go. Groynes will prevent the Beach-stream delivering sediments to the beach along its length. The beach however can grow back on the seawall. Later on, only the boulders that stick out need be removed. But how do we restore the Beach-stream?

What stopped the stream was not the T-head or the berthing face of South Jetty, but its approach. Therefore, the caissons of approach alone need to be removed and replaced by a bridge. That will allow smooth flow of the sediment-laden Beach-stream beneath during the ensuing monsoon. That should have restored the Beach-stream, but no more! Many more obstructions have come up downstream of South Jetty. All those must go too. In other words, the entire shoreline from the South Jetty right up to the beach must be made smooth, obstruction-free, just as Sir Robert Bristow had designed. Only then can the Beach-stream be restored to ensure sediment supply to Fort Kochi Beach. The beach will not only grow back, but also remain stable, as before.

There are other advantages of restoring the Beach-stream, besides getting back a beautiful beach. Siltation around South Jetty, Kochi Shipyard and Vallarpadam Terminal will go down, significantly. Consequently, we can expect siltation in approach channel to go down. Sediment supply to Puthuvypin will also go down. But that may not matter anymore, because the bunkering terminal would arrest whatever little that could have gone across. Therefore, the erosion of Puthuvypin is inescapable!

The International Bunkering Terminal on Puthuvypin also stands exposed to winds and waves. Fuelling the vessels during monsoon may prove difficult, probably dangerous too. Apparently, there is a plan to build an island breakwater to shelter the terminal. There is not enough room to fit one between terminal and channel. Therefore, it will have to come up south of the channel. But that may be too far to offer any kind of protection, unless they intend to make it very long.

Such a long offshore breakwater would no doubt affect the coast nearby. How and to what extent would depend on the design, length, position and orientation. With the current understanding of wave dynamics and modelling methods, things could go easily and seriously wrong. The outcomes can only be speculated.

Puthuvypin and Vallarpadam are not the only shelter-less projects on the West Coast. Navy's Third Naval Base, INS Kadamba, too has no shelter from the salt-laden monsoon winds.

CHAPTER 33

WHY KARWAR?

From the dawn of independence, the fledgling Indian Navy faced huge challenges—to defend a long coast and numerous far-flung islands, to deal with a belligerent neighbourhood, to guard the abundant wealth the surrounding seas beheld, and above all to secure the commanding geostrategic position that India naturally enjoyed being at the head of Indian Ocean. Navy therefore had to be equipped to face these challenges with warships and submarines. Warships and submarines need tranquil naval bases on the coast. Naval bases thus became the vital adjuncts to an emergent Navy.

At the time of independence, Navy only had two bases, one at Bombay (now Mumbai) and the other at Cochin (now Kochi). The fleet operated from Mumbai. Kochi handled onshore training. Both bases shared space with rapidly developing commercial ports. Therefore, it was only matter of time before Navy would be stifled. Navy's top brass realised this, even before the independence. So they felt the need for an exclusive naval base on the western seaboard. It came to be called the Third Naval Base or TNB. The choice, since the beginning, was Karwar.

Soon after the independence, Navy began to make ambitious plans to build the TNB at Karwar. TNB was elaborately discussed in the Annual

Senior Officers Conferences that followed the independence. Most probably, the 1962 Chinese aggression and conflicts with Pakistan thereafter forced the shelving of TNB. These border skirmishes led to a northerly shift in India's strategic thinking. Instead of aiming to be a naval power, naturally endowed with the coastline longer than the land border, the vast surrounding seas abounding in natural resources and the commanding geostrategic position at the head of Indian Ocean, India chose to become a local military power.

Forgotten were the oft-quoted wise words of Sir W Raleigh:

"For whosoever commands the sea commands the trade; whosoever commands the trade of the world commands the riches of the world, and consequently the world itself."[7]

Much later, in the early 1980s, Navy revived the plans to build the TNB. Before we go into what happened thereafter, let us examine why Karwar was in the first place chosen as the site for TNB.

The entire Indian Fleet, on many occasions, both before and after the independence, had anchored within the confines of Karwar Bay for rest and recreation or to ride out storms. The bay was roomy, deep and sheltered. But were these the only grounds for choosing Karwar? Probably yes! But there were other reasons just as important.

For a better appreciation of the choice, we must look beyond Karwar, in fact, along the entire West Coast, for a suitable site for setting up the TNB. Only then can we know if there could have been a better choice.

Monsoon waves render the West Coast impassable for nearly a third of the year, from mid-May to mid-September. Shelter is available only in the lee of headlands or within creeks and estuaries. There were but few such sites on this coast. Major ports have already come up there, but for Karwar. Mumbai and Jawaharlal Nehru ports are within the sheltered Mumbai or Thane creek. Mormugao Port is in the estuary of Zuari River, sheltered by a headland, Mormugao Head. Kochi Port is inside the estuary of Vembanad. The exception is New Mangalore Port, which was created more out of commercial expediency than prudent norms of locating an all-weather port.

---

[7]From the Works of Sir Walter Raleigh—A Discourse of the Invention of Ships, Anchors, Compass, &c

But even this port is not right on the coast to face the fury of monsoon waves, but within an artificial lagoon. We have seen how this port was made tranquil.

The ports that lacked shelter from the monsoon winds and waves stayed minor. They shut down operations during the monsoon, from mid-May to mid-September, every year.

The need for shelter from both wind and waves is far more critical for a naval base than for an 'all-weather' commercial port. The merchant vessels are strong-hulled to bear heavy loads of cargo. They can withstand some movement alongside. Moreover, they stay alongside only briefly. They discharge the cargo or load up and move on quickly. On the other hand, the warships are fragile-hulled to keep their weight low, for a better power-to-weight ratio. They must be able to move fast and have better endurance for the small load of fuel they can carry. Without adequate shelter, movement alongside the wharf would damage their fragile hulls.

Also unlike the merchant vessels, the warships must stay alongside for long durations. They are meant to standby to meet emergencies. So they are manned at all times. Without adequate shelter, the movement alongside would make life uncomfortable for the crew. That bears adversely on their fighting potential. The weapons do not fire on their own; the men behind matter more. They need to be comfortable and adequately rested alongside to remain alert at sea.

Besides, as the warships bang against wharfs or each other, their weapons and sensors are likely to be misaligned. The sensors, radars or sonars, detect the targets, which then point or guide the weapons, guns, missiles or torpedoes, towards the targets. For an accurate hit, weapons and sensors must be in perfect alignment. Misalignments, even slight, may result in the weapons missing targets. That may be disastrous in a battle. Re-aligning them is an elaborate process, and can be done only in absolute tranquil conditions.

As we have already noted, the salt-laden monsoon winds are extremely corrosive. Navy's assets, both floating and fixed, are prone to corrosion damage. The assets are expensive, hard to repair and harder still to replace. On the West Coast, a naval base therefore can come up only where there was adequate shelter from salt-laden monsoon winds.

A breakwater cannot provide such a shelter. It may stop the waves, but not the winds. Therefore, the shelter must necessarily be natural, either in the lee of a headland or deep within a creek or estuary. Karwar Bay was, at that time, the only unexploited site on the West Coast that was naturally sheltered, by a massive headland—Karwar Head. The bay therefore stayed tranquil throughout the year. And it was large enough to meet the need of a growing Navy.

Karwar, in spite of being so well sheltered, did not develop into a major port, because of poor hinterland connectivity. It is hemmed by high and densely wooded hills to east and south and Kalinadhi River to the north. Besides, there were no commercial or industrial centres near Karwar, because of the surrounding rugged terrain. That however provided the exclusivity and security that a naval base actually needed!

Besides being physically secure landward by the natural barriers, the Karwar Bay is just as secure from the seaward. Besides the towering Karwar Head to the south, it is fringed by number of islands. The Oyster Rock group of islands is to the west. Shimisgudda and Kurmagadgudda Islands are to the north. To the south is the Mogeragudda or Elephant Island. See the Figure 33.1 below. These islands can be turned into sentinel-posts to guard the bay from any clandestine intrusions from sea. Such a naturally secured base assumes immense significance nowadays, with sea-borne terror strikes becoming a reality.

A naval base also needs a sizeable flatland contiguous to the waterfront, much more than that for a commercial port. Besides the operational, maintenance, logistic, administrative and training facilities, the base needs room for housing, both for the personnel and their families. The base must necessarily include all civic amenities—schools, playfields, crèches, garages, parks, markets, cinemas, clubs, places of worship and more. It is essentially a self-contained township.

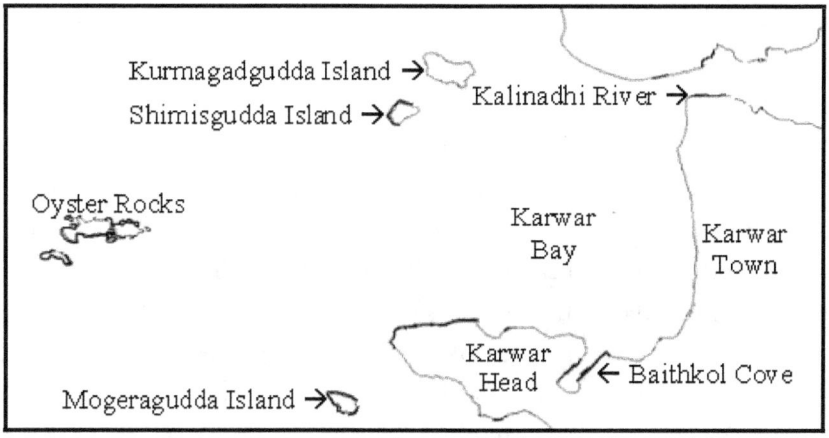

Figure 33.1: Karwar Bay and Town

Navy pegged the land requirement at about 8,000 acres. That incidentally was the available flatland at Karwar contiguous to the bay. So much flatland, right next to a sheltered bay, is not available anywhere else on the West Coast. There is flatland around Mangalore, but no sheltered waterfront. Elsewhere, the hills literally broached the sea.

Naval bases at Mumbai, Kochi and Vishakhapatnam have only a single channel each, both for entry and exit, used both by the warships and merchant vessels. Such a channel can be easily sabotaged. In this connection, I must share with you an incident, not of sabotage, but an accident, that took place in 1983 at Mumbai Harbour.

Many witnessed the sinking of a merchant ship, MV Cheri Chentek, within the channel, close to the harbour mouth. Master and crew were rescued, unhurt. Because the ship sank in the channel, the port had to be closed. INS Darshak was asked to search and locate the wreck and mark its position.

We surveyed a large area around the datum, the most probable position of the wreck, where the ship was seen going down. We sounded closely a large area around the datum and beyond the channel. We swept the seabed within that area using side scan sonars. There was no 'unlooked' seabed all around the datum. But we did not find the wreck. It seemed to have vanished into thin air or into some mysterious Bermuda Triangle. No one could tell where the wreck was. Without knowing where the wreck was, the port could not risk letting ships in or out. Another ship running into the wreck would be even more devastating for the port.

The search had gone on for nearly two weeks. Keeping a major port like Mumbai closed for so long was beginning to tell on the nation's economic health. Several ships were diverted. Perishable cargo aboard many others was lost. Matter became so serious that the Prime Minister's office stepped in to monitor the turn of events. The Chief Hydrographer to the Government of India, who heads the Navy's Hydrographic Surveying Branch, flew down to Mumbai to oversee the search. We, the hydrographers aboard INS Darshak, were under immense pressure, with so much attention focussed on us. Otherwise, seldom anyone bothered about us or what we did for a living, except sail the seas, with unflinching confidence, using the navigational charts that we so rigorously strive to make.

Having spent nearly a fortnight surveying, without a break, from the first light to last, often beyond, I was convinced that there was no wreck anywhere near the harbour mouth. But then, where did it go? It was no small boat, but a huge freighter. Only then did it dawn to us to talk to the ship's master about the state of his ship at the time of sinking. He was in the police lock-up. He knew only Greek, with a smattering of English.

With much difficulty, we learnt the ship's engines were running as the ship went down. Her holds were empty and battened down. Therefore, there was plenty of air trapped within the ship to provide the buoyancy. Most probably, the ship that everyone saw sinking did not hit the seabed, but remained afloat underwater, like a submarine. With engines running and sufficient air to keep it running for some time, there was enough motive power to take the ship far out to sea before sinking as the water seeped slowly into the holds and engine room. After much deliberation, the theory of the wreck floating out underwater was accepted. Many similar case-studies too were examined. The port was finally opened for traffic without finding the wreck.

What if someone deliberately sank a vessel in a single-channel naval base, unseen? Even a rumour may be sufficient to create chaos. Warships and submarines would remain bottled up until the channel is cleared. Therefore, for a naval base, having multiple channels for entry and exit is no luxury. On the West Coast, only Karwar provided that advantage.

The choice of Karwar is therefore hardly surprising. In fact, it was the best possible site on the West Coast. Probably, a gazette notification was issued earmarking Karwar for Navy, to prevent any encroachments or other developments taking place there. Those days it was just a coastal village, with about 5,000 inhabitants, who were mainly engaged in subsistence farming and traditional fishing. It should have remained so, until Navy chose to build the TNB, whenever. In fact, it did remain so until 1975. Then something sinister happened, not at Karwar, but nearby, to south, at tiny village by name Kokra.

Before I go into that, I must tell you how I got interested in the TNB project or Project Seabird.

CHAPTER 34

CASE OF THE SUNKEN PONTOON

Sometime in early 1983, while onboard INS Darshak, I was sent on temporary duty to Karwar, with orders to report to INS Sandhayak, which was surveying the Project Seabird site. She was required to return to her base port, Vishakhapatnam, because she was programmed for a foreign cruise. They expected some coastlining to remain incomplete. I was required to complete that. I reported to ship the following day, around noon, after a tiresome bus journey. She was berthed alongside at Baithkol Cove. See the bottom right corner of the Figure 33.1 above. There I was informed that I had no job and could return to Bombay. In the meantime, I received instructions from my ship to carry out the coastlining of Baithkol Cove before returning.

About a year ago, before I joined Darshak, she had surveyed Karwar Bay and its approaches. Baithkol Cove was not coastlined because of the wharf construction. Ironically, the survey was done at the behest of Government of Karnataka. The state government had prepared a Master Plan for turning Karwar into a major hub for commercial shipping, with wharfs, drydocks and ship repair yards. I also collected a copy of the Master

Plan from the local administration for my ship.

At that time, I knew nothing about the Project Seabird, let alone that Karwar was originally earmarked for the TNB. I doubt if anyone onboard Darshak knew that either. Anyway, the Karwar of 1983 was not a coastal village, but a bustling port town. The port was busy. Salt and furnace oil were being shipped in. Chemicals and granite were being shipped out.

On the following morning, as I was about to commence the coastlining of Baithkol Cove, a group of fishermen encircled me, menacingly, to be precise, 'gheraoed' me. They began to raise slogans against the Project Seabird. I was taken aback, again to be precise, scared. Fortunately, I knew the local language, Kannada. I spoke, as calmly as I could, to the one who I thought was the leader. I told him that I had nothing to do with the project. I was only doing the survey for the port. That was the truth anyway. That somehow calmed the situation.

Just then the police got there. Fishermen soon dispersed, quietly. The leader went across to a nearby trawler and returned with a small basket full of large 'tiger' prawns that would have cost a fortune at Bombay. He gave me the basket, probably to make amends. I accepted the gift and had it sent to the Circuit House, where I was staying.

Later that evening, the cook at Circuit House prepared the prawns in the local fare. It was remarkable. I had few people over, whom I had got to know during the day. Unlike the agitating fishermen, they were all enthusiastic about the project. They wanted to know more about it, but I knew nothing.

Though the newspapers talked about it, within the Navy, it was a 'hush-hush' affair, a highly classified matter. Information was only on need-to-know basis. There was no need for me to know, so I did not get to know anything, even to satisfy my curiosity. Going out of the way to satisfy one's curiosity on a classified matter was not a prudent thing to do. So I did not.

Two years later, in 1985, during the Long Hydrographic Course attachment to CWPRS, we learnt they were designing the Project Seabird. They had built a hydraulic model to evolve and test the design. We asked our guides to take us to the model, but were informed that it was too 'classified'.

Later that day, on the way back to the hostel, we walked past a large modern hangar, with 'Top Secret' emblazoned on its electrically operated doors. The hangar held the model of Project Seabird. None of us pressed for any information. Nor were our guides forthcoming. Probably, it was classified for them too.

In the rough and tumble of the hydrographer's life, I forgot about Project Seabird. But six years after the Long Hydrographic Course, at the Naval Dockyard Mumbai, my curiosity was again roused.

As the Officer-in-Charge ASD (B) Survey Unit I had to attend daily the ops meeting chaired by the C of Y. The meeting held in a small room above the Boat Pool. One day, in October 1991, soon after the meeting, I stopped for a chat with our naval constructor, the one who looked after the hull maintenance of dockyard's boats, tugs and passenger craft, including our survey motor boats. He had returned only that morning after a week's tour of temporary duty to Karwar. He had gone there, with a team of shipwright artificers, to inspect and write-off a pontoon that sank at the Project Seabird site.

Pontoon is a box-shaped floating platform made of steel. It is normally used as a buffer between a vessel and wharf or between two vessels berthed alongside, to prevent chafing. Pontoons therefore are also fitted with rubber fenders all around, and come in different sizes to suit vessels. The one that sank at Karwar was a large one, about ten metres by twelve metres. Such a pontoon is normally used for berthing an aircraft carrier. But at Karwar it served as a platform for oceanographic observations. It was moored offshore, but close to the coast, where it got corroded, holed and sank. It was hardly a sensational matter!

In any case, I was not concerned with the pontoon. I only wanted to enquire about his trip to Karwar and latest developments there. In his hand was the 'draft' writing-off report. During the conversation, for no particular reason, I took the report and casually glanced through. It was not a classified document.

The cardinal dates of the pontoon's life, boldly typed on the cover page, drew my attention. It was built in 1987. It was being written-off in 1991. That meant barely four years of life! That struck me as rather odd, because the pontoons lasted much longer, often more than twenty five years, that too, with little maintenance. Probably, it was made of some low grade steel or had no sacrificial anodes. Sacrificial anodes prevent corrosion to a great extent.

Out of curiosity, I asked him for more details, not wanting to read through the report. He told me that the dockyard had built the pontoon to better than standard specifications, because it was meant for the high profile Project Seabird. It had the required number of sacrificial anodes. Also, it was given several coats of anti-corrosive and anti-fouling paints. That made the matter more intriguing. Why then did the pontoon corrode so rapidly? He had no answer. I became concerned.

The site where pontoon corroded would soon be a naval base, where expensive steel-hulled warships and submarines would be based. If there was something in the region that caused the corrosion, the matter warranted immediate investigation and resolution.

The following day, during the ops meeting, I raised the matter with the C of Y. He assured me that he would look into it and do the needful. I do not know what transpired thereafter. The pontoon was however written-off without much ado. The report was duly filed. But I remained curious to know the truth.

After the brief stint at the dockyard, I was transferred to the Hydrographic School as an instructor. At the school, I had little time to spare to investigate the pontoon's corrosion, even though I was closer to Karwar, only about hundred kilometres away. In 1995, I took over the duties of Chief Instructor. I only got busier.

In December 1996, the Headquarters Goa Naval Area organised a whaler expedition to Karwar. Whaler is a 27-foot open boat, usually rowed with a bank of oars either side. It can also be rigged for sailing, with three sails—fore, main and mizzen. Hydrographic School was required to field a whaler in the expedition. I went along with a team of trainees. I enjoyed sailing the whaler more than any other sailboat, that too, out at sea. We set out early in the morning from NAD Jetty Chicalim. After an uneventful voyage, in calm seas and light winds, we reached Karwar that evening around sunset. We unrigged and secured the whalers at the inner end of Baithkol Cove, where the fishing trawlers were berthed.

It was quite late to return to Goa, so we had to spend the night at Karwar. I went over to the officer's mess of Project Seabird, which at that time was only a row of portable cabins at Baithkol Bay, south of Karwar Head. In the anteroom there was the sand model of Project Seabird, for briefing important visitors to the site. It was not classified.

The operational area or basin was coming up at the shelter-less Binaga Bay, between two small promontories, Binaga Point to the north and Arge Cape to the south. See the Figure 34.1 below. Anjadip Island was to the west, about two kilometres offshore. The island offered no shelter to the bay.

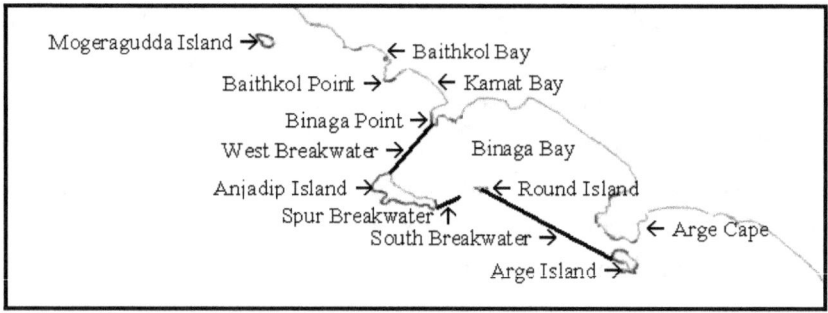

Figure 34.1: Project Seabird at Binaga Bay

On the model, a breakwater connected Binaga Point and Anjadip Island. Let us call it the west breakwater. A tiny breakwater extended from the island's southern tip, known as the spur breakwater. About a kilometre east of the island was a tiny rock islet, the Round Island.

The basin's main entrance was between spur breakwater and Round Island. About 3½ km south-east of Round Island was Arge Island. The south breakwater spanned Round and Arge islands. Other entrance to the basin was due south, about 500 m wide, between Arge Island and Cape. Heights of the breakwaters were 4½ m above mean sea level. That was indicated on the model.

From the officers present there I learnt that the officer's mess, club, and also the married officers' apartments were due to come up right on the seafront at the Kamat Bay, north of the Binaga Bay.

The estimated cost of the project then was about ₹ 50 billion. The figure, I believe, has been revised upwards to ₹ 100 billion, perhaps even more.

On the following day, early in the morning, I took a bus back to Goa. The whalers were towed back by a seaward defence boat.

CHAPTER 35

SEABIRD MODEL

In early 1997, I made a trip to Pune, to the CWPRS, for a meeting with their training division, to revise syllabus for the attachments of future Long Hydrographic Courses. After the meeting, only out of curiosity, I enquired about the Seabird Model that I was not allowed to see in 1985.

The model was still in place, but declassified, so they had no objection about my seeing it. They were planning to dismantle it anyway. By then they had ceased to be involved with the project. RADISSON of Australia and NEDECO of Netherlands were the new consultants for the project's marine works.

I had known one of the coastal engineers. We met first during the Karanja seismic survey, in 1983. We stayed in touch, occasionally though, sharing notes on ongoing coastal projects and problems. He agreed to accompany me to the model. He too was once associated with the project.

The huge hangar seemed bare. The model was too small compared to the size of the hangar. It was a tidal-cum-wave model, almost an exact replica of the sand model. My guide turned on the model. Simulated monsoon waves from the model's southwest corner began to lash the south breakwater, almost head on. The model also simultaneously simulated tides, but that mattered little, because the tide range at Karwar was only about a metre or so. The model did not appear too vertically exaggerated.

The south breakwater stopped the simulated monsoon waves. That made the basin tranquil. Little wave refractions crept through main entrance. The spur breakwater ensured that. The west breakwater mattered little to the tranquillity. I wondered why it was there in the first place. No waves came from the north anytime. Anjadip Island and Karwar Head ensured that. Probably, that was included only to enclose the basin, to make it secure.

My guide, pointing to the model's tranquil basin, said, with some pride, "You see, our design is perfect. The basin is both tranquil and siltation free. But we are no longer with the project. They have appointed some foreigners, as if they were better at designing on our coast. I wonder what their design is like."

The model basin was tranquil. I could see that. But it being siltation-free was only an assumption based on theory of littoral currents. If the breakwater stopped the waves, it would stop the littoral currents too. So the question of the currents delivering sediments into the basin would not arise. Therefore, the basin would not silt. That probably was the logic that he had applied.

I replied, "The basic design being executed is exactly as in this model, with no deviations."

He appeared pleased.

I continued, "But there will be little tranquillity in the basin, particularly during the monsoon. Besides, the basin would silt."

He wanted to know why.

I pointed to the south breakwater and said, "This breakwater is only 4½ metres above the mean sea level. It may stop the waves, but not the winds. Winds will blow over and wreck the tranquillity the breakwater is trying to achieve. Vessels berthed alongside will have a rough time during the monsoon. Perhaps, if you had tested the model in a wind tunnel, you may have seen the wind effect on the basin."

He replied, "I get the point, but there are no wind tunnel hydraulic models."

I said, "I know, but I think it may be time to go for one."

He then asked, "How do you say the basin would silt? Our studies did not indicate that."

I pointed to the shoreline of Binaga Bay on the model, and said, "Even though you haven't got it on your model, there are three big streams and many smaller ones that drain into the bay, but only during the monsoon. After the monsoon, these streams dry up and are barely discernible from rest of the landscape. But as the monsoon starts, these streams deliver huge amount of sediment-laden runoff into the bay. The sediments would normally have gone to sea, down the seabed slope. But now, the south breakwater would trap all that within the basin. That is siltation according to me."

He appeared confused. I had no time to make things any clearer. I had to leave.

As we walked back he pondered aloud, "Anyway, it does not matter now. We are out of the project. We will soon be dismantling the model. There is no point in wasting space."

I had nothing to say to that. Incidentally, the hangar and the model were built at considerable cost to the Project Seabird.

Their design was not wrong, per se, according to prevalent coastal theories. Notwithstanding, that probably was the best they could have done given a shelter-less site, like the Binaga Bay, to create a naval base. But why was the project coming up in that open bay, instead of the sheltered Karwar Bay?

When Navy staked claim for Karwar in early 1980s, it was no longer a small coastal village, but a bustling port town, with more than 50,000 inhabitants. Karwar's growth from village to town began with the development of its port. Until 1975, it was only a minor port, with little activity, but for an occasional coaster or dhow calling to embark rice or dry fish. Since then, it began to ship in huge quantities of salt and furnace oil. Why did a small coastal village require so much salt and furnace oil?

CHAPTER 36

MERCURY CONNECTION

The Centre for Science and Environment, a non-governmental organisation form New Delhi, had done a study of the Indian caustic-chlorine sector for its environmental impact, the Green Rating Project. Following is an excerpt from their report published in 2002[8].

Liquid Death

Death this time travelled through the waters and found its way into homes of innocent fishing folk in a seaside town of Japan, killing children in the wombs and affecting a number of people. Statistics cannot put an estimate to the suffering that spanned three decades.

The Chisso Corporation, one of the main employers of Minamata, was making petro-chemicals and plastics. From 1932 to 1968, Chisso Corporation dumped an estimated 27 tonnes of mercury compounds into the Minamata Bay. The destruction of large scale fishing areas following the dumping simply saw the exchange of money to buy

---

[8]Page 9, Green Rating Project—Environmental Rating of Indian Caustic-Chlorine Sector, Centre of Science and Environment

people off. The logic of the company was to pay people in exchange for polluting.

It was not till mid 1950s that people began to notice a strange phenomenon in animals and humans. People began to experience numbness in their limbs and lips. Their speech slurred and their vision constricted. Some people had serious brain damage. Birds started to drop dead from the skies. The valiant effort of a doctor from Chisso Corporation itself, Dr Hosokawa, brought the reasons for the disease to light. He faced resistance to his theory that linked the disease with the dumping of mercury compounds from the company into the sea. Chisso Corporation initially succeeded in buying the silence of the people but soon the incident came into national and international limelight. Though the victims testified at the United Nations Environmental Conference in Sweden, the UN did not intervene. Till a decade ago, the Japanese Courts were still resolving suitable compensation for the victims. It was Minamata which ultimately forced the Japanese government to ban mercury in all processes and products. It also heralded a new technology for caustic-chlorine industry – the membrane cell technology which gave it a new life.

Japanese ban of the mercury cell came in 1974. Other rich nations soon followed suit. But the demand for caustic soda and other chlorine products remained high as ever and booming, world over. The membrane cell technology was not yet in the design stage. It came years later. How did the world meet the demand in the interim?

The mercury cell based caustic-chlorine industry, dubbed dirty by rich nations, soon began to mushroom in poorer nations, where the governments were not too concerned about their citizens' health and safety. India was one of the destinations. Several caustic soda factories came up in India, mostly in the backward areas, probably under the guise of 'backward area development programmes'.

That may be how one came to be set up at Kokra, a typical backward area, in 1975, within a year of the Japanese ban. Kokra is a small village south of Karwar, about a kilometre as crow flies. If you took the National Highway NH 17, it is about 5 km from Karwar.

The factory was set up in a small flat cove, with hills on three sides and sea to the west, about a kilometre and half away. It went into production in 1975 with an installed capacity of 59,000 metric tonnes of caustic soda per annum. They manufactured caustic soda using mercury cell technology for 30 years. In 2005, they announced a switch to the mercury-free membrane cell technology.

In the mercury-based process, caustic soda is produced by a two-stage process. In the first stage, 25% salt solution or brine is electrolysed in an electrolyser, also known as mercury cell. The cell has an anode made of titanium-oxide, with a coating of ruthenium-oxide. The cathode is a layer of mercury, known as the mercury bath, at the bottom of the cell. That is how the cell gets its name.

When electric current is applied to the cell between anode and mercury bath, brine, chemically sodium chloride, is electrolysed into sodium and chlorine. Chlorine is a gas. It is removed from top of the cell and used for making host of other chemicals. Sodium is absorbed into the mercury bath below to form an amalgam of sodium and mercury. The depleted brine is removed from one end of the cell for reuse after adding more salt. From the other end, the amalgam is led into another chamber, the decomposer. That completes the stage 1.

In the stage 2, water is sprayed on the amalgam in the decomposer. Water reacts effervescently with sodium to form sodium hydroxide or caustic soda, the end-product. The reaction releases hydrogen, which may or may not be salvaged. With sodium thus removed, mercury is returned to the electrolyser. That completes one production cycle.

Even though it appears to be a closed-loop process, some mercury is lost in every cycle. Some goes out with the caustic soda and ends up in the market. The quantity however may be negligible. Electrolysis is a hot process, enough to turn some mercury into vapour, which gets extracted with the chlorine. That eventually enters the environment somewhere down the line. The quantity again may be quite small. Some mercury is lost through the water used to clean the mercury cell and decomposer periodically. The loss is significant. The bulk mercury loss is through the brine-sludge, the semi-solid residue left behind in the electrolyser. That accounts for almost two-third of the total mercury loss.

The factory's average mercury loss was put at about 151.6 grams for every tonne of caustic soda produced. You must not be surprised if the actual loss was higher. No one reports the mercury loss correctly. Ideally, from an environmental perspective, the mercury loss must be worked out on the basis of inventory transactions, that is, mercury bought, consumed and held in stock, instead of only the production losses. The net loss of mercury therefore can be worked out accurately from the store ledgers. That will include all types of losses—process and handling losses, accidental spills and pilferages. All that mercury, one way or the other, enters the environment. Such a system of accounting ought to be applied to every toxic chemical that the industries use.

For the present, let us accept the above figure and round that off to 150 grams per tonne of caustic soda produced. Two-third of this, that is, 100 grams of mercury per tonne of caustic soda produced is lost through the brine-sludge. That may not seem alarming, but it actually is!

The factory's average annual capacity utilisation was about 83%. In other words, they have been producing around 50,000 tonnes of caustic soda every year, for 30 years using, mercury cell technology. Therefore, the mercury loss through the brine-sludge alone computes to 150 tonnes in that period.

Now compare that with the 27 tonnes that Chisso Corporation released into Minamata Bay over a period of 36 years. The release of 27 tonnes of mercury caused many deaths and much misery at Minamata, which created so much public outcry that went up to the courts and later to United Nations, whereas the release of 150 tonnes caused none at all at Kokra! That is terribly intriguing. Indians are not immune to mercury poisoning. Or were the waste disposal arrangements so very efficient that no mercury entered the local environment?

The old Chart 2008, Karwar Harbour and Approaches, showed a pipeline, about four kilometres long, leading from the Kamat Bay into sea. Many senior naval officers, particularly those directly involved with the Project Seabird believed that the factory disposed their chemical waste to sea through this pipeline, after necessary stabilisation, whatever that meant.

If there was such a pipeline on the seabed, which they used to dispose chemical waste, then none should have been near the coast. The pontoon moored near the coast did however corrode and sink. Without corrosive chemicals near the coast that could not have happened. So the chemicals did reach the coast!

I sent a letter to DG Seabird, the Director General Project Seabird, expressing my concerns about chemicals in the waters there. Like the pontoon, even the warships and submarines that would soon be based there would be prone to corrosion damage, with disastrous consequences. Navy had already lost a warship at sea due to suspected corrosion damage. I lost a friend in that accident.

The seabed off Kamat Bay was also designated as 'sand-borrow' area, from where the sediments were to be excavated for use in the reclamation within the operational basin at Binaga Bay. The corrosive chemicals on the seabed would affect the steel used in the construction of foundations and other structures, with long-term consequences.

Quite unexpectedly, I received a prompt reply from DG Seabird allaying my fears about corrosion. His letter mentioned that the corrosion levels at the project site were comparable to that at Kochi and Mumbai.

That was impossible! Pontoons did not perish in four years there. I had to find the truth, at least, to satisfy my curiosity. Probably, the DG's inference may have been based on actual field-data, the analysis of water samples.

A small creek discharged at Kamat Bay. It has no name on the chart. Let us call it Kamat Creek. This creek leads right up to the factory. Near the factory, it is no more than a drain that one could easily leap across. If the chemicals did reach the coast through the Kamat Creek, somehow, we can infer it could not have happened during the non-monsoon months, because the creek discharged little water. The creek however did discharge copiously during the monsoon.

Therefore, if the chemicals did reach the coast through the creek, it would have happened only during the monsoon. By the end of monsoon, it would stop. Soon whatever chemicals that came through the creek would disperse or sink into the seabed. Therefore, the water samples collected during the non-monsoon months may have little chemicals to show. Incidentally, water sampling is impossible during the monsoon, because of the rough sea.

The puzzle began to solve itself. I had to visit factory to see where they stored the brine-sludge and other chemical waste, even if it was only a temporary storage. The environmental laws do allow the temporary storage of toxic waste within the factory premises pending 'permanent' disposal.

CHAPTER 37

CURIOUS CASE OF THE BOTTLED WATER

Taking a day's leave in December 1997, I made a quiet visit to Karwar, accompanied by a fellow hydrographer, in his car. By then, I was transferred to the Headquarters Goa Naval Area as Staff Officer (Quartering and Adventure Sports).

At Karwar, we met one of the hydrographers attached with the Project Seabird. He agreed to take us to the factory. He knew some of the managers there. He also told us that he had been there a number of times and it was prim and proper. According to him, they had excellent certifications, particularly with regards to waste disposal. I did not say anything to that.

Before setting out to factory, he took us home for lunch. At lunch, he served us 'bottled' water. I felt that strange. And as we were leaving, a 20-litre plastic bottle of water was delivered to the doorstep. I could not contain my curiosity.

I asked him why he was using expensive bottled water for drinking and what was wrong with the local water supply. He told me that he used bottled water not only for drinking, but also for cooking. Water supply to his house came from a bore-well. His landowner had advised him not use it for drinking as it may be contaminated with human faeces, because of the number of leaky septic tanks around. Faeces or was it something else, I wondered. Pathogens can be destroyed by boiling. So why use expensive bottled water? He then mentioned most people he knew used bottled water not only for drinking and cooking, but also for bathing their children. That certainly was strange!

Soon after that, we went to the factory. Near the main gate, to the left, was an open sandy field fringed by trees, which could have easily passed off as a playfield. We stopped there. Our escort got off the car to have a word with someone he met there. For no particular reason, I kept looking towards the playfield, wondering what I would see inside the factory. I was not expecting anyone to be playing there. It was not the time.

Then I saw a tractor go into the field. It backed towards one side, jacked up its tipper, dumped whatever it was carrying, and drove away. It was over in minutes. Curiosity got better of me. I got off the car to take a closer look. I had no doubt what the tractor had dumped. It was brine-sludge!

So that was where they 'stored' the brine-sludge—in an open sandy playfield! I went back to the car. A short while later, our escort too returned. I told him that I had seen whatever I had to, so there was no point in going into the factory. I did not tell him what I saw. He did not ask. He then took us to the project site, where the construction was going on in full swing. Before I tell you what happened there, I must tell you what happens to the brine-sludge dumped on that sandy playfield.

The factory generated about 63 kg of brine-sludge for every tonne of caustic soda they produced. Therefore, every year they generated about 3,150 tonnes of brine-sludge. If all that was dumped into that 'playfield' no bigger than 200 m by 250 m for so many years, there should have been a hillock of brine-sludge. But there was no hillock there, only a flat grassless field. Not a blade of grass grows there in stark contrast to the verdant surroundings. Where did all the brine-sludge go?

There are no secure landfills anywhere around the factory to store the brine-sludge. Also, there are no secure lagoons within the factory's premises to store the mercury-laced wastewater. Landfills and lagoons are impossible there, or for that matter, anywhere in that region. That was also the reason why there was no hillock of brine-sludge.

The reason is monsoon rains. The region receives more than 3,000 mm rainfall every year. If that is not enough, lot more water flows as runoff down the surrounding hills through the region, literally, seeping through the sandy soil. With so much water flowing, no landfill or lagoon can be secure enough to contain chemical waste—liquid, semi-solid or solid, mercury laced or otherwise.

Every monsoon, the water from rain and runoff fluidised the brine-sludge and other chemical waste dumped on that playfield, sending a portion to sea through Kamat Creek into Kamat Bay, and rest into groundwater aquifers.

Some years ago a team from the National Institute of Oceanography conducted a survey of Kalinadhi River, flowing north of Karwar, to measure levels of toxic chemicals in the water and on the bed. They were expecting to find some arsenic. Paper mills were known to discharge arsenic, and there was one upstream.

Instead, they found unusually high levels of mercury. They checked the upstream for any possible source, but found none. They were mystified. One of the scientists, who took part in the survey, was with me during the XVIII Indian Expedition to Antarctica (1998-99). He told me about the survey. I asked him if the groundwater aquifers of the surrounding regions were connected to the river.

"*Most likely yes*" was his answer.

Kalinadhi flowed less than ten kilometres from the factory. I told him how the mercury may have reached the river. The well-to-do at Karwar were aware of mercury leaching into the groundwater. Most of them were associated with factory, one way or the other. So they used the bottled water. But no one talked about it.

Chemicals reached the coast during the early days of monsoon. Soon there was little left on the playfield. By the time monsoon ended, whatever that reached the coast either dispersed or settled down onto the seabed. Therefore, post-monsoon water samples had little to show.

The pontoon was moored near the coast or beached during monsoon, therefore stayed in contact with chemicals throughout. It corroded rapidly, in spite of being made of good steel, with several coats of anti-corrosive paints and sacrificial anodes. It was holed and sank. That was all there to it.

What about people in the locality, were they not affected? After all, they used the mercury-laced groundwater for just about everything, from cooking to washing, to water their cattle and to irrigate their small landholdings to grow rice and vegetables. I had no means to investigate that. I did however get some indications of the Minamata disease in the region.

At the project site, our escort got talking to one of the contractors in his native tongue, Telugu, the language spoken in Andhra Pradesh. I was introduced to him. I learnt from him that most workers employed at site were from Andhra Pradesh. There were no locals. That was strange too. Normally, locals were given preference for employment at such projects.

The contractor was quick to tell me that locals were lazy. They were not interested in any sort of work, much less the heavy, tiresome construction labour, despite being poor and under-employed. He added they would rather starve than work.

One of the early symptoms of mercury poisoning is chronic lethargy, a complete lack of energy and enthusiasm. Nerve disorders, paralysis and death came later, much slowly.

Later, at Bangalore, in December 2000, during a conference on lakes and wetlands organised by the Centre for Ecological Sciences, Indian Institute of Science, I met a scientist from the centre, who some years ago had taken part in an Environmental Impact Assessment survey for Project Seabird. Surprisingly, he was unaware of mercury pollution in the area. He was not even aware of the existence of a caustic soda factory close to the study site. Apparently, the factory was not included in the survey. They only had to survey the flora and fauna in the designated area!

I asked the scientist, "Are not the humans part of local fauna?"

He looked at me as though I said something stupid.

So I rephrased my question, "What about the local population? Did you notice anything strange about them?"

His replied, "Strange? No Sir, they were plain lazy, in fact, very lazy. They wouldn't care to help us even with our baggage. They just sat there and stared at us."

People were lazy and unenthusiastic about work or livelihood, because they were being poisoned by mercury, which they were consuming through water, vegetables, rice, milk and fish. Even at Minamata, people realised that they were being poisoned only when they were found to be lacking in energy and enthusiasm even after consuming the energy-filled fish caught in the Minamata Bay. The fish were poisoned by mercury.

I wonder how many in the region may have died, after prolonged suffering, which included paralysis. From some unconfirmed sources I learnt that there were many cases of paralysis in the region. The poor locals, mostly illiterate, accepted all that as matter of fate, without knowing that they were being poisoned.

It is not quite possible that the local administration would not to have known about it. After all, the people must have gone to the local health units or hospitals. What about the doctors? Was there was not a single Hosokawa among them?

For those suffering from mercury poisoning, it may have been 'Bhopal' played out in agonisingly slow motion. But Bhopal was an accident, Kokra was not!

CHAPTER 38

FUTURE OF KADAMBA

To produce 50,000 tonnes of caustic soda a year, the factory needed about 70,000 tonnes of salt. The salt came from Singach in Gujarat by ships to Karwar. Besides salt, the factory needed electric power for the electrolysis. It consumed about 3000 units of power for producing a tonne of caustic soda, which means annually 15 million units! All that power was generated in-house, with four generators fuelled by furnace oil. The oil probably came from one of the refineries at Mumbai, again, by ships.

The ships also began to call at Karwar to take away chemicals that the factory produced. The Karwar Port began to develop. Ancillary industries and allied businesses came up at Karwar, where there was plenty of flatland available. No one thought of Navy's TNB.

Employment opportunities soared. More people moved in. Shops and markets came up to meet the demands of the surging population. Homes and offices were built. Schools and colleges came up, and hospitals too. The local administration therefore had to be spruced up. It became the district headquarters. Karwar became a boom town!

In the early 1980s, when Navy staked claim for Karwar, the state government was in a fix. It became a political hot potato, too hot to handle. Navy was offered alternate site anywhere on the Karnataka Coast. The state did not want to lose TNB.

Navy, once again starting out from scratch, began to scout for a new site. Many places were examined—Binaga Bay, Ratnagiri, Pawas Bay, Goa, Tadri, Mangalore and Tuticorin. Hydrographic surveys of the shortlisted sites were undertaken. I had taken part in some without knowing the real purpose.

Finally, for reasons not quite clear, Navy chose Binaga Bay for the TNB. They staked claim for 12,000 acres of land, but finally settled for 8,000 acres. That incidentally was the flatland available at Karwar contiguous to the sheltered Karwar Bay. Instead, Navy got 8,000 acres spread along 28 km of shelter-less coast, on hill tops, in forests and God knows where. Navy now bears a huge burden of guarding all that land needlessly.

Saddest cut was for the people who were evicted. They were not meant to be, whereas the well-heeled economic migrants at Karwar stayed on.

CWPRS was tasked to design the operational basin at the shelter-less Binaga Bay. The foreign consultants were later hired to design and build the infrastructure within the basin. They did nothing to the basic design that CWPRS had prepared. Even if they were to design the base, they would not have come up with anything different. The prevalent understanding of coastal processes was the same everywhere.

It was obvious that they knew little about the monsoon dynamics on the West Coast. They could do no work during the monsoon. The project was executed only during the non-monsoon months! No wonder it took so long and cost so much. The taxpayers picked up the bill. They had no choice. Few knew about the project anyway. And those who knew did not probably know how wrong things were going.

Getting to know the Project Seabird was worse than opening the proverbial can of worms or Pandora's Box. You may choose the worst. I conveyed my findings to some of the senior officers, who I thought would be concerned, through a demi-official letter, in 1999. Not one acknowledged the letter.

Though the day-to-day affairs were handled by a senior naval officer, the DG Seabird, it was a Cabinet Sub-committee that called the shots. In any case, the project was too far advanced for anyone to do anything.

My letter did have some effect. I found myself under transfer to Karwar for the project. At my seniority, there was little I could do to alter the course of events or design. In any case, there were two hydrographers, much senior to me, already with the project. Surely, they knew better.

I sought a cancellation. It was granted, fortunately. Thus I came to Bangalore in April 2000. TNB was commissioned on 31 May 2005 as INS Kadamba, though plenty of work still remains. What may be its future? That may not be difficult to envisage.

During the monsoon, there is little tranquillity within the operational basin. Conditions are terrible during storms. There are no easy solutions for that.

Basin also silts. That can be managed by maintenance dredging. But there was an attempt to prevent the sediment-laden runoff from the adjoining hills flowing into the basin by diverting that elsewhere. That is fraught with risks. The runoff flowing in is huge. Diverting that elsewhere would lead to hinterland floods, with loss of life and property. That probably did happen!

A problem that has no solution is corrosion. Almost everything at the base is exposed to the salt-laden monsoon winds. Kadamba is an example of how a township ought not to be built on the West Coast. We need not go far to see how one ought to be built!

Karwar Town is an excellent example. Karwar Head protects the town from the corrosive monsoon winds. Even then, there is almost nothing in the town that is exposed directly to the winds from sea in any season. Immediately landward of the 'surf-less' Karwar Beach is a swath of trees, casuarinas and palms. The National Highway NH 17 runs landward of this swath of trees. Landward of the highway are the public offices, each with a sizeable garden in front. Behind the offices are the schools, play grounds and parks. The town's main road is after that. East of the main road is rest of the town.

Such an arrangement is impossible at Kadamba, because there is little flatland between the hills and coast. The town planners engaged to design Kadamba obviously had no knowledge of the monsoon. They probably felt that living on the seafront would be elegant. The officers' apartments are right on the coast at the Kamat Bay. We must ask them how they feel about it, particularly during the monsoon.

One of the important principles of war is 'defence in depth'. Kadamba has none, when it comes to engaging its principal enemy, the salt-laden monsoon winds. The base may be reeling under the impact, with enough damage already. There is little anybody can do, anyway. In the years to come, the base will become un-maintainable.

When it comes to physical security, less said the better. The base is the cynosure for every vehicle passing on the highway above!

There is yet another threat that looms large over the base, the threat of a cyclone. I wonder if that has seriously been considered. A cyclone can be devastating on a shelter-less coast, which also happens to be built up and densely inhabited.

The threat is an alarming 'two' per cent. That means two cyclones can strike the coast in a span of hundred years. Since 1891, when the record of cyclones has been kept, two have struck the coast. Both incidentally were classed severe or super cyclones. The loss of life and destruction of property then may have been probably minimal due to the low population density in the region. That no longer is the case today, at least, around the Binaga Bay. It has been a while since a cyclone has struck the region. Global warming is only going to make things worse, particularly the storm's fury. It is good to be prepared. But that is easier said than done. Evacuating the base may be the only way to save lives. What about the infrastructure? A storm surge trapped between hills on one side and a long breakwater on the other can be devastating.

Imagine all that happened, because a Japanese corporation dumped 27 tonnes of mercury into a small bay, the Minamata Bay, from 1932 to 1968. As a result, many Japanese fell sick and some died. And so the mercury poisoning came to be called Minamata disease. But for that, Japan would not have banned the mercury cell. The technology would not have come to India, at least, so cheaply. The caustic soda factory then would not have been set up at Kokra, where there was neither salt nor power. But then, why did the factory came up there? I have no answer.

For fulfilling its role effectively, Navy must get back Karwar Bay to build a new base on the contiguous land, the Karwar Town, and forgo Binaga Bay. That may sound preposterous, more so, after having spent more than ₹ 100 billion. There is no other choice. It was a mistake. It is time we accepted that and moved on. The nation with such a long and dynamic coast also needs to stake more research effort to understand it.

With that, let us go back in time, to about 5,000 years ago, to the Indus Valley Civilisation.

CHAPTER 39

END OF A CIVILISATION

The Indian Subcontinent was home of one of the oldest civilisations, the Indus Valley Civilisation. The civilisation flourished on the banks of Indus River about 5,000 years ago. Like the civilisations of Mesopotamia and Egypt, this one too was based on flood-plain agriculture on either bank of the Indus River.

Mohenjo-Daro, one of its principal settlements, was situated nearly 500 km upstream of the Indus Delta. Mesopotamian seals at Mohenjo-Daro and Indus seals at Mesopotamia bear testimonies to the trade between the two civilisations. It is also believed that the trade was by sea, via the Persian Gulf. If that was so, where was the seaport of the Indus Valley Civilisation?

There is no evidence of any port within the present Indus Delta. In fact, the delta is hardly a place for a seaport, because of shallow depths, meandering channels, shifting sandbanks and frequent floods.

Lothal is a rather nondescript place about 80 km south of Gandhinagar, the capital of Gujarat. It is however quite famous as one of the places linked to the Indus Valley Civilisation. The site was excavated in mid-1950s by the Archaeological Survey of India. The most remarkable feature of the excavations at Lothal is the dock. The slightly trapezoidal dock is about 210 m long (north-south) and 35 m wide (east-west). It is difficult to say how deep the dock was. When I visited Lothal in July 2009, it was almost entirely silted, with some rainwater pooled at one corner.

The dock made little sense at Lothal, because there is no navigable waterway nearby. The nearest one, Sabarmati River, is about 15 km away. The archaeologists, who excavated the site, however believed that the Sabarmati those days flowed either near or through Lothal, later meandered away. So they concluded Lothal was a river port with a dock. That is portrayed in the artist's impression of Lothal in its heydays, which they had prepared. You can find it in most books on ancient Indian history and at the Internet sites dealing with Lothal, including the Wikipedia. The dock, as per the artist's impression, was connected to the river by a natural stream or artificial canal. Through the river, Lothal was presumably connected to Gulf of Khambhat, thence to sea.

If there was sea trade between Indus Valley and Mesopotamian civilisations, then the large sea-going ships must have called at Lothal. That means the river or some other canal from the head of Gulf of Khambhat to Lothal must have been quite deep too. A closer examination of the region's topography and the flood plain of Sabarmati however shows that the nearest the river could have come to Lothal was about 12 km, any time in the past. That was the maximum meandering the river could have undergone from its present course. A river can only meander within its flood plain, which for Sabarmati was narrow, because of the surrounding terrain.

The dock's design also pointed to its use only in the tidal waters. Such a dock is seldom found on the bank of a non-tidal waterway or river. If Sabarmati or some other river did flow through Lothal, so far away from the gulf, it would have been almost certainly non-tidal.

But to suggest Lothal was a seaport in tidal waters also does not seem quite tenable. It is today about 25 km inland, north-northwest of Gulf of Khambhat. On the other hand, if it was a seaport, then 5,000 years ago the Gulf of Khambhat did extend up to the dock. That means the gulf has shrunk since then. Is there any evidence in support?

On the gulf's west coast, about 30 km south of Lothal, was a famous port, Port Dholera. It finds mention even in the latest edition of West Coast of India Pilot, the nautical publication that the mariners use to navigate along the West Coast. The pilot can be seen as a kind of coastal travelogue for mariners. There is another one for the East Coast, the Bay of Bengal Pilot. There are about 70 pilots, covering every coast and ocean in the world.

Few centuries back, Port Dholera used to be 'one of the chief cotton marts in Gulf of Khambhat'[9] and a thriving port. British East India Company is known to have handled cotton trade through the port. Accretion of the gulf's west coast put Dholera inland. For some time, the trade went on through Bhadar Creek that connected Dholera to gulf. Later, because of the creek silting, goods were stopped about 3 km from its mouth, at Whittle Bandar. From there goods were taken to Dholera by road. Further silting of the creek put an end to all trade with Dholera. Today, there is no Port Dholera, but only a landlocked Dholera Village, more than ten kilometres inland from the gulf's west coast!

We can therefore deduce that 5,000 years ago Gulf of Khambhat did extend more to the west and north, that is, up to Lothal. Those days, Sabarmati, like Mahisagar, Dhadhar and Narmada today, must have drained into the gulf from east, about ten kilometres north of its present mouth. So the gulf had shrunk from the west, in other words, its west coast accreted. We have seen the process of accretion of Puthuvypin. That was wave driven. But within the gulf, there are practically no waves. How then did the gulf's west coast accrete?

To understand the process, we must examine the landscape along the gulf's west coast, from the low to high water line, and also landward. Unlike most other coasts, here the low and high water lines are far apart, often kilometres apart, because of the high tide range in the gulf and also the gentle gradient of coast. The shallow gradient continues deep inland. On the other hand, the waterfront, which can be seen only at the lowest tide, during the springs, is a stark contrast to the adjoining coast and landscape. It is a sheer drop, a vertical cliff of black, hard clay. There is no cliff of clay

---

[9]Article 8.181, West Coast of India Pilot, 2003 Edition

across the mouths of the shallow creeks that drain into the gulf from west. The sediments discharged by the creeks spread across their mouths like a fan, rendering them shallow, and with a gentle gradient into the gulf.

Everywhere else the waterfront is cliff of clay, which generally stands about a metre or so above the water level at the lowest tide. Along this cliffy waterfront, for about a hundred meters or so landward, the soil is entirely clay. But as we go landward, we find a layer of silt and some fine sand over the clay base. Further inland the layer of sand on surface gets thicker and sand progressively gets coarser. As we go more inland, the surface layer is coarse sand and gravel, with the finer sediments below over a clay base.

The process of accretion works like this. Whenever it rains on the eastern side of the Kathiawar Peninsula, occasionally though, sediment-laden runoff flows towards the gulf, but slowly, because of the gentle gradient of the landform. As the runoff flows gently downslope, the coarser sediments, gravel and sand, deposit upstream or landward. As it flows further downstream towards the gulf, finer sediments, fine sand and silt, get deposited. What eventually reaches the waterfront is clay, which normally is viscous fluid clay. That is very sticky. As the sticky fluid clay flows across the waterfront, it does not rush down, but slithers down slowly, sticking to it, giving the vertical waterfront a coat of clay. The coast thus accretes by the clay sticking to the waterfront layer by layer. The process may appear exceedingly slow, but it is interminable. So the gulf's west coast will go on accreting. In other words, the gulf will go on shrinking, eventually to become a narrow waterway only to drain the four rivers flowing into it.

Therefore, the inference that the gulf extended up to Lothal, about 5,000 years ago, does seem quite reasonable. But that still does not automatically make Lothal a good enough seaport. All settlements of the civilisation, at least, those that have been excavated, are far north and widely separated from each other. Surely, they would not have relied on lumbering ox-drawn carts to ferry goods to Lothal and back. A water transport system would have been more appropriate. That means there must have been a navigable waterway that extended northward from Lothal.

The topography does point to that. It is clear that such a waterway extended north up to Little Rann. Again, if we discount the siltation, we can deduce that Little Rann was a deep lagoon, not a salt marsh of today. Most probably, those days, Little Rann and Rann of Khachch may have been a single large lagoon, into which drained a branch of Indus, or perhaps the entire Indus. Therefore, not only the lagoon and waterway up to Lothal navigable, but the water in them was fresh.

The settlements south of Indus, so far excavated, were around this lagoon. There may be more waiting to be found, even along the waterway leading up to Lothal. Therefore, Lothal was connected to all settlements, right up to Mohenjo-Daro and beyond as far north as Indus was navigable those days. Lothal may have been the only seaport for trade with the outside world catering to entire civilisation. The principal, perhaps the only mode of transport between settlements must have been small single-mast sailboats. Some clay seals depict images of such sailboats. Probably, the seals were fares for travel or to transport goods.

Siltation at the head of gulf, because of sediments brought down through the waterway, must have necessitated the dock. Large sea-going vessels used for trade with outside world would have entered or left the dock during the high tide. When the tide was low, the vessels remained in the dock, unloading or loading. From the dock, small sailboats carried goods north to various settlements, and also brought goods from there to embark the vessels bound for distant shores.

There is nothing to suggest that the dock had a lock-gate. Therefore, it must have been quite deep, so that there was always sufficient depth to berth large vessels, even when tide was low. Such a deep dock, without a lock-gate, was sure to silt. That may have necessitated regular maintenance dredging to restore the depths. They must have had some primitive form of dredging, probably, workmen going down to dock bottom and coming up with sediments in baskets. Similar technique is still employed at many places on the subcontinent to mine sand from the rivers.

By desilting the dock, completely, we may be able to recover its original bed, which I presume too may be brick laid, like the sides. Besides, there may be remnants of ships and other nautical artefacts on the bed. It may be worth the effort to desilt the dock completely, if it had not been done during the initial excavations. That may shed more light on the dock's design and on the civilisation in general.

Therefore, both the lagoon and waterway were vital to all settlements south of Indus, both for transport and freshwater. Besides, the entire civilisation needed the seaport at Lothal for trade with the outside world. There was no other. The siltation rendered the lagoon shallow and then split it into two. With that Indus ceased to flow through the lagoon. It then may not have taken long for the lagoons to turn into salt marshes—Little Rann and Rann of Khachch.

Accretion and siltation together put an end to the waterway up to Lothal. Without freshwater, means of inter-settlement transport and trade with outside world, the complex and isolated settlements could not survive. The people therefore migrated elsewhere, seeking greener pastures. Continuing accretion of the gulf's west coast then put Lothal deep inland. Therefore, it was not an invading army or culture or some esoteric climate change that brought down Indus Valley Civilization, but natural processes of accretion and siltation.

It is possible to trace this ancient waterway from Lothal to Little Rann. Remnants exist today as wetlands, lakes, salt marshes and lowlands. It is about hundred kilometres long.

It may also be possible to restore this waterway to connect the Gulf of Khambhat to Little Rann and thence to the Gulf of Khachch through the Hansthal Creek. This waterway can then be made deep enough for modern ships. It could be led further north through the Rann of Khachch to sail the vessels closer to the northern heartlands of India, and probably even linked up to the Indus to the west.

But today, such a navigable waterway may be only wishful thinking, because plans are seriously afoot to build a dam across Gulf of Khambhat. That is the next story, the story of Kalpasar.

CHAPTER 40

FROM BRIDGE TO DAM

The Saurashtra region of Gujarat, west of the Gulf of Khambhat, is perennially drought prone. The last drought, in the 1970s, was indeed severe. For providing drinking water, the government ran water-trains from the water-surplus region east of the gulf to the drought-hit west. The trains had to go round the gulf to reach their destination. So at that time, an idea was mooted to build a bridge across the gulf, from Hansot in the east to Bhavnagar in the west. Such a bridge would reduce distance between the two regions from 250 to 70 km.

After the drought passed, the idea was however not followed up. Instead, the state's Irrigation Department built more than 50,000 check-dams, across almost every stream in the Saurashtra region to trap, literally, every drop of runoff that came from the infrequent rains. That, to a great extent, mitigated the drought-proneness of the region.

In the early 1990s, another idea was mooted, a novel one yet. Instead of a bridge, the proposal was to build a dam across the Gulf of Khambhat. Such a dam would trap the waters of four rivers—Narmada, Dhadhar, Mahisagar and Sabarmati. Everyone felt the water from these rivers was going wastefully to sea through the gulf.

Everyone believed that the dam would transform the drought-prone Saurashtra into an agriculturally and industrially productive region. Besides, it would bear the road and rail traffic, to give much fillip to the region's economic boom. The dam seemed like a vast improvement over the bridge. It would however turn the gulf into a lake, which came to be called the Kalpasar, meaning the Eternal Lake.

A Dutch consulting firm declared the idea of Kalpasar highly workable. So the Kalpasar Department was constituted to oversee the designing and construction of the dam and its associated canal network. The department was manned by dam and irrigation experts, both serving and retired personnel from the Irrigation Department, and supported by a panel of consultants, both Indian and foreign.

As per the initial design, the dam was to be in two parts, one for freshwater storage and the other for tidal power generation. It was about 64 km long and stretched from Ghogha in the west to Hansot in the east. Besides the mouths of four rivers, the dam also enclosed the ports Bhavnagar and Dahej.

The tidal power generation was included, because of the high range of tide in the gulf. The plans to generate power from tides were being explored long before the idea of Kalpasar. The tide range on the western side of the gulf is higher than on the east, therefore the tidal power plant came to be on the dam's western half. The eastern half was for the freshwater storage, because the rivers drained across the gulf's east coast.

But after detailed evaluation, the plan to generate power form the tides was dropped due to technical complexities and high cost vis-à-vis the advantages. The project was thereafter only aimed at freshwater storage, the Kalpasar. So the dam was shifted north, mainly to keep the ports Bhavnagar and Dahej out. That reduced its length to about 30 km across the gulf. That also put the Narmada Estuary out of the dam, but not its water. The water, whatever little that flowed out of the Sardar Sarovar Dam, would be trapped upstream of the estuary by a barrage and sent into the Kalpasar through a diversion canal. The barrage would therefore cut off the water, hence the sediment supply to the estuary downstream. You can guess the impact—interminable erosion!

From the beginning, one thing that bothered the experts was the dam's closure. The dam on a river is normally built simultaneously from both its banks. As the gap between the two sections narrows, the surge of water through becomes intense. That makes the closure extremely difficult. So to ease the surge, the water upstream is diverted downstream, temporarily, through a diversion canal. Besides, the closure is attempted only during the dry season, when little water flowed through the river.

For the Kalpasar Dam, the option for a diversion canal did not exist. Like a river, there is no valley here. The landform either side is flat, barely above the sea level, which made such a diversion canal, one side or the other, impracticable. Also, there is nothing called a dry season in the gulf. The water level in the gulf is decided not by inflow, but by the tide.

The experts however were aware of the swift tidal stream in the gulf, which they believed would cause the surge during closure, just as in the case of a river dam. But they were not sure how to deal with it. So the coastal experts were called in to study the closure problem being a matter of tides and tidal stream.

The coastal experts, most probably, studied the problem on a 'computerised' tidal model of the gulf. On the computer model, when it comes to simulating the tidal stream, the entire water column, from the surface down to the bed, flows one way or the other, as if powered by gravity, just like on a hydraulic model. Therefore, the rate of flow depended on the bed gradient, just as in a river. As a result, even on their model, the closure produced a surge.

At the coast, as we have already noted, the tidal stream is a surface phenomenon. When this surface flowing stream enters or leaves a narrow gulf or creek there is no surge. The surge can only happen when the water is flowing under gravity through a narrow opening, like a river flowing through a gorge. Because of build up of head upstream of the gorge, the water surges out through. For the river dam, it is gorge-like situation at the time of closure.

Come to think of it, the closure of Kalpasar Dam should hardly be a problem. No matter how swift the tidal stream may be, its peak rate lasts only for a short while. When the tide changes, one way or the other, for some time, the rate drops to almost zero, in other words, it becomes almost still water. That is when to close the dam. That again need not be done in a hurry, all in one go, but in stages, at every change of tide. Moreover, there is little flow, anytime, below the lowest tide level or the Chart Datum.

Nevertheless, the experts continued to be anxious about the surge that the swift tidal stream would cause at the time of closure. The swift tidal stream in the gulf is due to the abnormally high tide range. But why is it so abnormally high in the Gulf of Khambhat, unlike most other places on the coast?

CHAPTER 41

RESONATING TIDES

For a recap, tide range is the difference in the heights of successive high and low tides. It is a variable. But for any place on the coast, it is highest during the springs, the full and new moons. That is when the moon and sun are in a line, and so together exert maximum gravitational pull on the sea surface. The tide range is least during the neaps, half way between springs, when the moon and sun are at right angles to each other.

Besides the temporal variations due to the changes in the relative positions of sun and moon, the tide range also varies spatially, that is, from one place to another, even across the Gulf of Khambhat. The spring tide range at Bhavnagar is about 8.8 m, whereas just across, on the gulf's east coast, at Port Dahej, it is only about 7.4 m. At Alang, to the south, it is 6.2 m. Further south, at the entrance, on the western end, the Gopinath Point, it is only about 5 m, whereas on the eastern end, the Suvali Point or Hazira, it is 5.7 m. No tidal data is currently available north of Bhavnagar.

Many are under the impression that the high tide range in gulf, consequently the swift tidal stream was due to 'funnelling' water through

217

'inverted funnel-shaped' landform towards the gulf's head. Incidentally, there are other places with similar 'inverted funnel-shaped' landforms, but with no high tide ranges, for example, Mumbai Harbour. The tide range at Mumbai is neither so high as in the gulf nor is there any significant variation across the harbour. Actually, there is nothing called 'funnelling' effect, when it comes to tides or tidal stream.

The real reason for the phenomenal tides in the Gulf of Khambhat is a simple phenomenon known as tidal resonance. Even if you are not familiar with the term 'resonance', you may have experienced it, sometime or the other, usually in a city bus. When the bus stops, with its engine idling, sometimes, it begins to vibrate, suddenly. That is resonance. That happens when the frequency or rate of vibrations of the idling engine matches the 'natural frequency' of the bus. Every object, natural or manmade, has a natural frequency. It is the rate at which the body naturally vibrates or resonates, but with little provocation.

As long as the frequency of the idling engine matches the natural frequency of the bus, it will continue to resonate. The moment the driver put his foot on accelerator pedal and changes the engine's settings, even slightly, the bus will cease to resonate.

Oceans, seas and gulfs too have their respective natural frequencies, which depend on their shape, size, depth and bottom configuration. The natural frequency of a water body influences its response to the tidal force or the net gravitational pull exerted by moon and sun combined.

The tidal force causes the tide to rise and fall, but acts varyingly on the sea surface, because of the earth's rotation about its axis and relative changes in the positions of moon and sun. That in turn produces a 'wave' of tide that spans the earth. This wave of tide is very complex, changing its shape from one place to the other, and also with time.

But this complex wave can be resolved or broken up into number of regular waves, each with a simple form, known as the tidal harmonics. Each tidal harmonics is based on some motion of earth, moon or sun, or a combination thereof. There are more than seventy tidal harmonics, which when combined make up the actual, complex wave of tide. The harmonics vary from place to place. They can be derived for any place on the coast or even for an entire ocean, sea or gulf, mathematically, by analysing the tidal data covering a long period, ideally, a tidal epoch or 19½ years. The harmonics are then used to derive the predicted tides for each place. If the harmonics have been accurately derived, the tidal predictions for that place would also be accurate.

The tide in Gulf of Khambhat is semi-diurnal. That means the tide in the gulf changes every twelve hours. The abnormally high tide range is, most probably, due to the gulf, in its present form, resonating with the moon's semi-diurnal harmonics, which incidentally happens to be the strongest of all harmonics. This can be empirically verified.

The phenomenon of tidal resonance in the gulf may have serious bearing on the outcome of Kalpasar.

CHAPTER 42

FUTURE OF KALPASAR

The water in Gulf of Khambhat is salty, with the salinity varying between 2.8% to 3.2%. The experts have estimated that in about four to five years after building the dam, the impounded water would turn fresh, or nearly so, because of the inflow of freshwater from the rivers and the outflow across dam's spillways, as it fills up each time.

All around the gulf are saltpans. A wide swath along the west coast, north of Bhavnagar, is one of the most productive salt making regions in the world. Unlike most other salt producing areas, here the water is not drawn from the gulf for solar evaporation into salt. Instead, it is pumped up from open wells or groundwater aquifers. Water in these aquifers is up to twenty times saltier than the gulf. In other words, salinity in the aquifers varies between 30% and 60%, whereas in the gulf it is a meagre 3%. It is brine in the groundwater aquifers along the gulf's west coast!

After the dam is built, these brine aquifers will go underwater. The salt from the aquifers will no doubt leach into the reservoir. No one can be sure how fast the leaching would take place and how long it would go on. The amount of salt held in the aquifers is probably not yet assessed. But it could be substantial. Therefore, in all probabilities, Kalpasar may remain an eternal 'salt' lake. Even it staying 'eternal' may be in question.

The lifespan of dam has been pegged at 250 years. It is the duration up to which the dam can effectively store water, before being fully silted up. Silting of a dam is an ongoing process, with sediments being constantly delivered into it by the river. Kalpasar Dam has four rivers bringing sediments, besides what comes with the runoff flowing in from west through the numerous creeks and also with the overland runoff. We also must not forget that the gulf's west coast is also accreting. That means we are dealing with an interminably shrinking gulf.

Most probably, the dam's lifespan estimate may have been based on siltation in river dams in the region. The river dam usually has sluice gates to release water at will, besides the water being continuously released for power generation. The dam must release some water always to maintain a minimum flow downstream. During the floods or when it rains heavily in the catchments, the water released is much more. Through all these discharges, a huge load of sediments leave the dam. That brings down the net rate of siltation.

Kalpasar Dam, on the other hand, only has spillways to release water. That is water literally overflowing the dam, whenever it fills up. The dense sediments will therefore remain within the dam, on the reservoir bed. Therefore, the actual rate of siltation within Kalpasar will be much higher. In other words, the dam will have a much shorter lifespan, at best, few decades.

Besides whatever happens within the dam, it may be what happens on its downstream that may be more worrisome. That can have more widespread, long term and unredeemable consequences. The dam, with only spillways to discharge water, will completely cut off the sediment supply to the coast downstream, both sides. The erosion downstream is therefore a foregone conclusion.

On the east coast, it will surely go on up to the mouth of Tapi River, the next source of sediments to the coast. That will destroy the ports Dahej and Hazira, and the adjoining landscape. The erosion around Hazira may prove hazardous, because of the tonnes of oil trapped in the tanks and pipelines. On the west, the entire stretch up to Gopinath Point will erode, and soon be reduced to steep cliffs.

If the Sardar Sarovar Dam, so deep inland, could set off erosion within the Narmada Estuary, you can imagine what the Kalpasar Dam can do sitting right on the coast.

Erosion may not be the only problem downstream. The dam will significantly alter the gulf's shape and size. With that the natural frequency of the gulf or what is left of it downstream will change. The gulf may no longer resonate as before. The real possibility will be a lower tide range downstream. That in turn would reduce the width of intertidal zone, the area between low and high water lines. Impact on the flora and fauna that thrive in the zone will be devastating. Moreover, the zone is made up of mudflats that are largely clayey. When the zone's width reduces, the mudflats would soon dry up and turn into a dusty landscape. Dry clay is fine dust. Dust storms that ensue will exasperate the already deteriorating environmental situation. Dust storms are both frequent and devastating, where the large lakes and inland seas have shrunk, exposing their beds. Such dust storms can affect the human and animal health over a very large area around the gulf.

The reduction in tide range will also affect the ports and fishing harbours in the entire region southward. They will no longer be able to handle large vessels, which currently they do taking advantage of the high tide range. The ports Bhavnagar, Dahej and Hazira, whatever salvageable after the erosion, will therefore be rendered useless. The ship breaking industry at Alang, which relies entirely on the high tide range to bring vessels onshore, will be consigned to history. The fishing community obviously will be the worst hit.

On the other hand, if the tide range increases, though unlikely, the problems may be no less devastating. The intertidal zone would get wider. That would put large tracts of built-up and cultivated areas under seawater.

There may be other deleterious effects that are yet unknown. There is little knowhow at present about the impact of tidal regime change. Kalpasar may prove to be an environmental nightmare. Risks are enormous and hardly understood.

Water is a necessity, but at what price? The dams are known to create more problems than they solve. Kalpasar Dam may not be an exception. Problems that this dam throws up may be on a scale yet unimagined. The project is expected to cost ₹ 2 trillion or more. Coupled with the enormous environmental cost, that may be too huge a price to pay for water.

Like the Kalpasar, yet another water project of mega proportions seems to have caught everyone's imagination, the inter-linking of rivers. The concept is simple. Link up the rivers through long canals, so that the surplus water from one region can be delivered to those deficient. The proponents claim the project would solve nation's water woes, besides putting an end to the perennial floods. Obviously, they have not considered the coast, which also happens to be an integral part of the nation.

When the rivers are linked up and their waters shared, therefore consumed, little would reach the coast. As a result, the sediment supply to the coast would go down drastically, setting off erosion, particularly around the river mouths. On the East Coast, the erosion would spread north of every such river mouth, because the longshore drift would become unsustainable. The erosion will go on interminably till the rocks inland are exposed to the waves. The permanent spits across the major sediment delivery zones, the river deltas, for example, the Godavari Sand Spit across the Kakinada Bay, would soon erode away.

With the sediment discharge of Ganges and all its branches going down, the delta, the Sunderbans, will steadily become scarce and eventually disappear. The impact may be inestimable, but beyond doubt catastrophic for the ecosystem, which cannot migrate anywhere, like the humans who subsist there.

Ultimately, it will be the unsuspecting coastal communities that end up paying the price. They would lose their homes, land and livelihood. They would then have to be displaced away from the coast, to alien settings. That would also put an end to their centuries old sea-dependent cultures and traditions. It is therefore clear that the proponents of such projects have little idea what a coast is and how it works. So, what is the coast anyway?

CHAPTER 43

COASTAL CONTINUUM

We have so far seen how the coast works and what happens when its natural working is disturbed. But we are yet to define coast. I trust by now we know the coast well enough to be able to coin a good definition. It must also be quite clear that the coast is about the movement of sediments, until it turns rocky due to erosion. And that only happens when the coast somehow stops getting sediment supply from land. We can therefore define the coast as *the zone of transfer of sediments from land to sea*. It is not an insipid border between land and sea.

But the coast cannot be viewed in isolation, as an entity existing by itself. Actually, there is nothing fixed about a coast, its form or its sediments or even its position between land and sea. It is actually part of a continuum. For want of a name, let us shall call it coastal continuum. This continuum has three parts—coastal plain, coast and continental shelf. It is a continuum only because these three parts are formed by a one-step process, which again is not anything esoteric, but simple deposition of sediments eroded from land and brought down to seabed, across the land margin, also known as the continental margin. That is to say, the coastal continuum, all its three parts, is made up of the sediments deposited on the seabed along the land, a part of which being above the sea surface. Therefore, it is several kilometres thick, tapering down to the seaward from land. I shall presently make these things clearer. The weight of sediments thus deposited on the seabed would compress those deep down into rocks, the sedimentary rocks.

Coast is the narrowest part of the continuum. Where it is on the continuum at any time is quite important, because its position is continually shifting, due to changes in sea level. The sea level change can be both short and long term. Short term changes are due to tides, and occasionally due to storm surges. Long term changes are due to changes in global temperature. As the global temperature drops, water freezes around the poles, thus bringing down sea level everywhere. During the last ice-age, about 20,000 years ago, coast, world over, must have been far seaward of where it is today. As the global temperature rises, the sea level too goes up with release of water frozen around the poles. The coast then creeps into the coastal plain.

The landward part of the continuum is the coastal plain. It is a gently sloping landform, normally uninfluenced by sea, except during storm surges and tsunamis. It differs from the adjoining land, being entirely composed of deposited sediments, all the way up from the seabed. That will be become clear, when we go through the process of coastal continuum formation.

The seaward part of the continuum is the continental shelf. It stays underwater. It is normally defined as the undersea extension of land. That gives an impression that the land somehow went underwater. But that may be so only in some exceptional cases, when some part of the land goes underwater due to prolonged erosion, for example, the bed of Palk Bay and some portion of the seabed south of Kanyakumari and also south of the Kathiawar Peninsula, or by sudden subsidence of land following a massive earthquake, for example, some parts of the Great Nicobar Island around the Indira Gandhi Point went underwater after the 2004 earthquake that caused a massive Tsunami. But in general, like the coastal plain, even the continental shelf is formed entirely by the deposit of land-eroded sediments.

When a large landmass begins to breakup, cracks form deep beneath the crust. Through these cracks, the magma or molten rock from the earth's interior pushes its way vertically up to the surface. That splits the landmass vertically from bottom upwards. The rising magma pushes apart the separated landmasses and begins to spread between, deep down, to form the new seafloor. Seawater rushes in to cool and harden the new seafloor. As the separated landmasses drift apart, the seafloor also spreads, as magma continues to spew out of the initial crack. The magma spewing out of the seabed crack eventually turns into an undersea volcanic ridge that may be hundreds or even thousands of kilometres long.

The seaward edge or margin of freshly separated landmasses would therefore be a vertical cliff of rock all the way down to the new seafloor, a sheer drop several kilometres down. Therefore, to start with, there is no coastal continuum along the land or continental margin.

Land has been eroding all the time, even before the breakup. The eroded sediments would however begin to flow down to sea across the margin only as it begins to rain. That is when the continuum begins to form, by the deposit of eroded sediments on the seabed along the margin. Therefore, greater the rainfall along the margin and landward, greater is the sediment delivery across the margin, and so the continuum forms faster. As the rains continue, the continuum gets wider. Therefore, width of the continuum is directly proportional to the rainfall over the adjoining landscape.

If the continuum is wide today, we can deduce that the adjoining landscape had been getting rainfall copiously for a long time. Say, if the adjoining region is dry today, but with a wide continuum, we can infer that the region did get heavy rains in the past. For example, the Kathiawar Peninsula gets very little rainfall today, but has a wide continuum. That means the peninsula did receive heavy rains in the past, and also for a long time.

The formation of the continuum, besides the amount of sediments delivered across the margin, also depends on the rate at which the sediment-laden runoff flows across. That in turn depends on the gradient of the landscape adjoining the margin. Let us consider two types of landform along the margin, one a flat, gently sloping tableland or plateau and the other a steep mountain side. That, more or less, was the situation on the Indian Landmass, when it broke away from Gondwanaland.

Where the adjoining landform has a gentle gradient, as it begins to rain, the sediment-laden runoff flows rather slowly towards the margin. The heavier sediments—pebbles, gravel and sand—gets deposited upstream, and so what reaches the margin is mainly fluid clay. The fluid clay is sticky. It does not rush down to the seabed, but rather spreads downward on the margin's vertical surface, like a coat of paint. As layer by layer of clay gets coated, the margin slowly accretes. Thus the coastal plain evolves through a process of accretion, just like the situation on the accreting west coast of Gulf of Khambhat. See the Chapter 39: End of a Civilisation.

The coastal plain remains above the sea level for quite some time, depending on how high the margin was above the sea. Like the preceding landform, the coastal plain too acquires a gentle slope seaward, whereas its seaward edge would remain steep, a sheer drop down to the seabed, like the original margin, but smoothened out by the layering of clay. As the accretion continues, the coastal plain dips underwater. The process of accretion does not stop, but continues underwater to form the continental shelf. The shelf's seaward edge would however remain steep for a long time to come. See the right side of the schematic diagram in the Figure 44.1 below.

On the other hand, if the landform adjoining the margin is a steep mountain side, as it begins to rain, the sediment-laden runoff rushes down to the seabed. Sediments thus start to pile up along the foot of the margin, on the seabed, like the bed silting up. The continental shelf thus begins to form first. As the sediments continue to pile up on the seabed, the shelf spreads seaward and so gets wider. As more sediments come down, portion of the shelf along the margin eventually rises above the sea surface to form a narrow coastal plain. See the left side of the schematic diagram in the Figure 44.1 below.

Generally, coast is the only zone for the transfer of sediments from land to sea. But in some exceptional cases, occasionally though, the sediments go to sea, literally, bypassing the coast. That is what the mystery of the dead coast is all about.

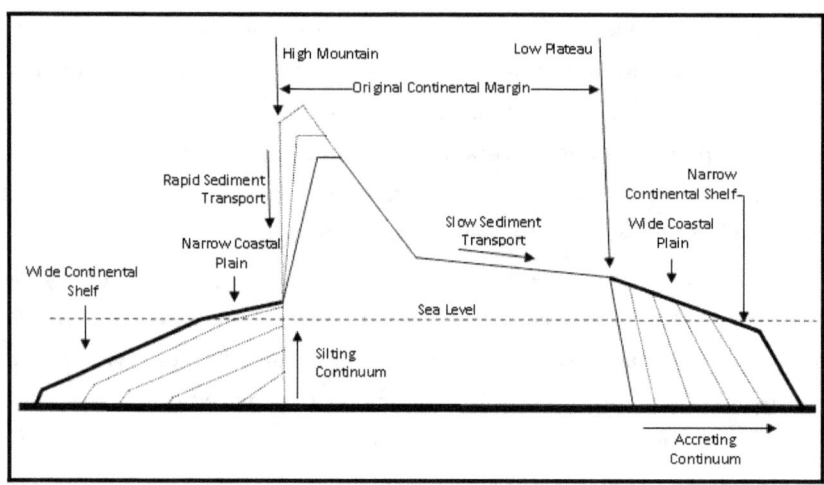

Figure 44.1: Formation of the Coastal Continuum

CHAPTER 44

MYSTERY OF THE DEAD COAST

During the thick of monsoon, when big waves are lashing the Kerala Coast, the traditional fisherfolk are shore bound. They have little to do. Some of them await the Chakara, which literally means 'dead coast'. That is when the wave-lashed coast suddenly turns calm, as if dead. It is perhaps unique to the Kerala Coast. Also, it does not occur everywhere, but only at few places, which are more or less fixed on the coastal stretches of Alapuzha, Ernakulam, Thrissur and Kozhikode.

The most famous Chakara is the one on Alapuzha Coast, at Purakkad, few kilometres south of the Alapuzha Town, the Venice of India. Late Thakazhi Sivasankaran Pillai, a popular Malayalam writer, epitomised Purakkad's Chakara in his novel Chemmeen. The novel was later made into a movie that went on to win the President's Gold Medal for the Best Film in 1965.

The popularity of Chakara is because of the abundant fish it brings. Along the 'dead coast', the water literally teems with fish. Folklores say that during Chakara, the fish can be scooped up in bamboo baskets, without even having to cast the net. When the news of Chakara breaks, the fisherfolk from the nearby areas rush to harvest the bounty it brings. They must hurry, because it is a short-lived phenomenon.

Over the centuries, Chakara attained a mythical status. Fisherfolk believe that it is a blessing conferred by the Kadalamma, the Sea Goddess or Mother, when venturing out to sea becomes difficult due to the rough monsoon waves. But in these recent times, it has become rather elusive. Probably, Kadalamma is upset. So they offer prayers and sacrifices to appease her, but to no avail. Before going into why it has become elusive, we must know how it actually occurs.

The experts have found that the occurrence of Chakara coincided with the sudden formation of offshore mudbanks. The waves then break on the mudbanks rendering the sea coastward suddenly calm. That is simple! But how do mudbanks form offshore so suddenly? Several theories have been proposed, but none supported by fieldwork, because no one took the trouble to go out to sea at that time. That is actually quite difficult. One had to wait in the rough monsoon sea, sometimes indefinitely, for the Chakara. That can deter even the hardiest of mariners.

Long ago, some hydrographers did venture into the rough monsoon sea to study the mudbanks. As the news of Chakara broke, they rushed out to look for the mudbanks off the Alapuzha Coast. Finding mudbanks was easy. Waves were breaking on them, as if on a shallow patch.

They took some soundings and collected samples of mud. That was sufficient to get to the bottom of Chakara. Mud was soft, dark green, and oily[10]. It was rich in terragenic biomass, plant and animal matter from land, including bits of roots and trunks. That means the mud came from land. How could the mud on land reach the seabed offshore to form the mudbanks?

That is possible only if it was pushed out from land under great pressure. Only a powerful hydrostatic pressure device can do that. But where on the land can so much hydrostatic pressure build up, powerful enough to push

---

[10]Article 5.59, West Coast of India Pilot, 2003 Edition

the mud out to sea, so forcefully, so copiously and so suddenly? Let us examine the region landward, where Chakara usually occurs, say, at Purakkad on the Alapuzha Coast.

Landward of Purakkad Beach is a shallow lagoon. This lagoon is incapable of generating so much hydrostatic pressure on its bed, even if it filled up to brim. Besides, it cannot fill up, because the water would seep out to sea through the sand. So this lagoon cannot be that pressure device. Let us go inland. The landscape inland is flat for kilometres around, criss-crossed by rivers, streams and canals, some draining into Vembanad to the north, while others into another large backwater to the south, the Ashtamudi Kayal. The flat landscape is known as the Kuttanad. It is one of the most fertile regions in Kerala. The soil is clayey and rich in biomass, which makes it fecund.

Few thousand years ago, there was no Kuttanad. Instead, it was a massive backwater, Vembanad and Ashtamudi Kayal combined. Kuttanad was the result of years of siltation. In other words, siltation split the massive backwater into two. Soon the surface layer of the silted portion dried out in sun, but below, it stayed soft, still fluid, fluid-mud, even today.

Therefore, two backwaters, though appear separate on the surface, are connected by a subterranean layer of fluid-mud. Between this fluid-mud layer and the sea is another layer of hard clay, which ordinarily is impermeable. The clay layer prevents the fresh water from land seeping out to sea. We have dealt with the phenomenon before, in the context the siltation of New Mangalore Port. See the Chapter 21: River through the Lagoon.

With the onset of monsoon rains, the runoff flows into the two backwaters through a number of rivers and streams and overland flows. Both these backwaters discharge to sea through narrow outlets—Kochi for Vembanad and Neendakara for Ashtamudi. But these outlets cannot drain the entire monsoon inflow fast enough. So the water level in the backwaters rises. That in turn increases the hydrostatic pressure on subterranean fluid-mud between the two, that is, below the Kuttanad. As the water level in the backwaters rises with continued inflow, the hydrostatic pressure exceeds a threshold value, whatever that may be. That is when the subterranean fluid-mud is pushed out to sea puncturing channels through the clay lining between land and sea, like squeezing out toothpaste.

The fluid-mud thus squeezed out erupts from the seabed offshore like mud-volcanoes to form mudbanks. The waves immediately start to break on them rendering the sea coastward calm. That is Chakara, the 'dead coast'.

The discharge of fluid-mud releases the hydrostatic pressure built up at the backwaters, either end. With that, the mud supply to seabed ceases, just as suddenly. The mud within the subterranean channels soon congeals, sealing them shut. The waves soon scour down the mudbanks and spread the mud on seabed. They again break on the coast. That brings to close the Chakara for the season.

The process is same everywhere on the Kerala Coast, where Chakara occurs. Landward is invariably a flat Kuttanad-type landscape with backwaters north and south.

Why is the calm sea during Chakara water teeming with fish? Nearshore fish population thrive on the nutrients coming from land. The coast that delivers little nutrients to sea usually has a small nearshore fish population. The fish population off the Australian Coast is small, because little nutrients flow out to sea from the desert.

The mudbanks, on the other hand, are rich in nutrients. Fish know that instinctively after years of conditioning. Long before the humans, the fish were waiting for Chakara. The nutrient rich mud that eventually settles down on the seabed becomes the spawning ground for several varieties of fish. It is hardly surprising that the monsoon is the spawning season for many varieties of fish off the Kerala Coast.

Why is Chakara becoming elusive in these recent years? For Chakara to occur, the water level in the two backwaters, either side of the flatland, must exceed a particular level. Only then will the hydrostatic pressure exceed the threshold level. If the inflow is insufficient, the level will not increase, because whatever little that flows in will be soon drained through the narrow openings on the coast. Then the hydrostatic pressure will be insufficient to push the fluid-mud out to sea.

Most rivers and streams draining into the backwaters have been dammed or barraged or their waters diverted for other uses. As a result, inflow into the backwaters has reduced significantly. There is now less chance of sharp rise in water level, unless it rains heavily right over the backwaters. Another factor is siltation. The reduced depths would also bring down the pressure build up on the bed when the backwaters fill up.

Chakara may soon be a thing of past, if not already. It may remain only in coastal ballads and folklores. Apparently, little can be done about it. It is a phenomenon in transit, like the barrier islands and coral reefs off West Kathiawar Coast. Probably, Chakara did occur there too, when monsoon rains filled those backwaters, before the coast broke up into barrier islands. Could that be the future of Kerala Coast too, now that it has begun to erode rampantly?

With that mystery unravelled, we can move on to the last one, the origin and evolution of Indian Coast. From the preceding stories, it must be quite clear to you that, one way or the other, the monsoon is behind the processes at work on the Indian Coast. Therefore, let us first get to know the phenomenon of monsoon, before examining how it may have influenced the evolution of this beautiful and dynamic coast.

CHAPTER 45

WIND BASICS

The Indian Subcontinent gets most of its rainfall from the monsoon. The word monsoon comes from the Hindi or Urdu word 'mousam', which means weather or season. It is the season of rain in the subcontinent. In fact, people wait for the monsoon only for the rains it brings. But monsoon is not so much about rains as it has been made out to be. It is more about winds. Rains are what the winds deliver on making landfall on the subcontinent, after blowing across many seas, for more than half the year, covering over eight thousand kilometres!

Like the waves, even the winds that make waves are poorly understood. The meteorologists will disagree. Nevertheless, before we go into the story of monsoon, which is only a seasonal wind system, we must know the winds better, perhaps little differently from what we think we know, that is, our preconceptions. For that, once again, we must ask some basic questions that a child would normally ask. That may be the only way to escape the clutches of our preconceptions.

So let us begin by asking the most basic question. What is wind? That is easy. Wind is air in motion. How does air get moving to become wind? Or, what makes the wind blow?

The familiar answer is—the wind blows from high pressure to low pressure. That is like saying that the wind blows where and when there is a pressure gradient. It may therefore seem that the wind blowing from high to low pressure is like the water flowing from high to low level. That may be why there is so much talk about pressure in our meteorological bulletins—high pressures, low pressures, troughs of low pressures, so on and so forth.

When the air gets blowing as wind, the Bernoulli's Law once more enters the scene, rather grandly. Recall the discussions on wave generation in the Chapter 10: Making Wave Sense. Anyway, for a recap, when the air starts to move, it gains in kinetic energy, so to conserve its total energy, the pressure it exerts on land or sea drops. In other words, the atmospheric pressure actually drops as the wind blows! We can easily confirm that in a wind tunnel, where the model of a prototype aircraft is tested.

When the air is blown on the wing of the prototype aircraft model in the wind tunnel, there is drop in pressure both above and below it. The shape of the wing above and below the wing is so designed that the drop in pressure above is more than that it is below. That makes the aircraft to take off and fly!

So, if the wind blowing is causing the drop in pressure, how can we say that it is the drop in pressure from one place to another that makes the wind blow? Pressure gradient therefore is not what makes the wind blow. That is only an effect! In other words, the change in atmosphere pressures anywhere on the earth is wind driven, invariably, not the other way. This may come as rude shock to many weather experts. So then, what is making the wind blow?

Instead, let us ask—what purpose does the wind blowing serve? As I mentioned before, nature does nothing without purpose. The purpose may hold the key to understanding the mystery!

Earth gets energy from sun in the form of heat radiations. But this energy supply from the sun is not uniform. Some places get more heat, while others less, sometimes quite consistently. So the answer to the above question is—the wind blows only to spread the heat around the earth, to counter the uneven heating. If the air remained still, earth would have been unliveable. Some places would be extremely hot, while the others unbearably cold. Besides the air in motion or wind, the ocean currents or the seawater in motion also spread the heat unevenly absorbed by the oceans.

Therefore, it is the difference in heat from one place to another that makes the wind blow. In other words, it is the heat or thermal gradient that drives the wind. Steeper the thermal gradient faster is the wind!

Besides uneven heating, the earth, depending on nature of the surface, absorbs heat unevenly. The capacity of sea to absorb heat is about ten times that of land. Sea holds the heat, whereas land heats up and quickly re-radiates the heat back to the air above or the atmosphere.

Fortunately, the atmosphere above does not heat up directly by the sun. If it did, little heat would have reached the earth's surface. That would have made the earth quite unliveable.

When the sun heats up a place, the air above heats up and rises. Everyone knows that. That is what makes the hot air balloons go up. The rising hot air draws the heat away from that place to the cooler atmosphere above. As the hot air rises, it makes room for the air from cooler regions nearby to flow in to get heated and rise. As the hot air rises, there is also a drop in pressure there.

The atmospheric pressure, we have noted before, depends on the number of air molecules in a parcel of air and their individual energies. When air on the surface heats up, the molecules gains energy and flies up. That leaves behind fewer molecules there, hence the drop in pressure. Soon the low-energy molecules from the cooler surroundings rush to gain energy. But that gets interpreted as the wind blowing into a low pressure. The purpose is only to draw away heat from that place and spread it to cooler atmosphere above. The low pressure is incidental. If after a while, sun stops heating the place, as it does, when its declination changes, then no more hot air would rise from there, and so no more colder air would be drawn in. The low pressure thus formed would disappear.

Sun's declination is the angle its apparent orbit makes with earth's axis of rotation. It is only apparent, because it is the earth's axis that is tilted by about 23½ degrees to its orbit around the sun. In a year, as it completes one revolution, the sun's declination changes from 23½ degrees north latitude or Tropic of Cancer to 23½ degrees south latitude or Tropic of Capricorn, and then back.

On 21 June every year, the apparent sun reaches the Tropic of Cancer and then starts to go south. On 21 Dec it reaches the Tropic of Capricorn and starts to go north. The sun therefore spends longer duration, continuously, around the two tropics, north and south, on its way up and then down. Therefore, the two tropics are hotter than anywhere else. As a result, most deserts are along the tropics.

When the sun's declination equals the latitude of a place, between the two tropics, it is peak summer there. The sun's declination therefore has much bearing on the earth's heating, consequently on winds that blow over the surface and also in the upper atmosphere. That in turn influences the changing weather patterns on the earth.

As the hot air from the earth's surface rises it would lose heat to cooler atmosphere above. As it cools, it would sink into some other warmer place nearby. As the cool air sinks into a particular zone steadily for some time, the pressure there increases. In other words, it becomes a high pressure zone. Also, it gets cooler there. That makes the air from the high pressure zone susceptible to blow to some other hot zone, which inevitably is at a lower pressure. That however gets interpreted as wind blowing from high pressure to low pressure.

There is yet another phenomenon that affects the wind direction. Because of the earth's rotation, wind does not follow a straight path, but curves. The curving of wind as it blows is known as the Coriolis Effect. The effect was first described by Gustave-Gaspard Coriolis in 1835, hence the name. He said that a body moving in a particular direction is deflected right, if it also happens to rotate counter-clockwise or is deflected left if it rotates clockwise. Right and left are reckoned with the direction the body is moving. Coriolis Effect therefore makes wind blowing in the northern hemisphere turn right, and left in the southern hemisphere.

As the wind blows over the sea, because of the pressure drop it creates on the surface, it also sucks up moisture. Faster the wind blows greater is the drop in pressure on the sea surface, hence more moisture it sucks up, and so wetter it gets. As a corollary, longer the wind traverses over the sea, wetter it gets. So you know why the monsoon is such a wet wind system. Incidentally, the heat absorbed by sea is drawn away with the moisture. That is the primary purpose of the wind blowing, to spread the heat.

Due to the earth's rotation and winds blowing nearly all the time, sun cannot possibly heat up any place excessively, particularly the sea, where even the seawater is flowing as currents, though not due to the effect of winds as many believe. Therefore, there is an upper limit to how high a place can heat up. If that was so, there must also be an upper limit to how fast winds can blow. Therefore, even the wind at sea has a speed limit, whatever that may be.

That means the extent the atmospheric pressure can drop on the sea surface, which is but an effect of wind blowing, also has an upper limit. Therefore, the waves and storm surges have natural height ceiling. I think the figure that Rear Admiral Sir Francis Beaufort worked out may be that ceiling, though he did not quite specify it so—14 to 16 m. Therefore, some of the old mariners' tales about encountering mountainous waves are fictional!

The biggest waves I have witnessed may have been about 12 m high. That is only an estimate. I had no means to measure that. That happened when INS Betwa, the old training ship, just about managed to escape the eye of a severe cyclone in May 1979. I was a cadet onboard. The ship was on a World Meteorological Organisation directed MONEX or monsoon exercise. She was let adrift in the Bay of Bengal, about 400 nautical miles southeast of Chennai, with a host of meteorologists and oceanographers aboard, who were measuring various weather and atmospheric parameters. But during the cyclone, only a few aboard were fit enough from sea sickness to witness the violent marvel of the cyclone. I was one of the lucky ones.

I had yet another encounter with a cyclone, in May 1990, but not anywhere nearby. That was when I was tracking the progress of a surveying ship, INS Sandhayak, which went into the eye of a severe cyclone. She was battered beyond recognition, and almost floundered, but had a providential escape!

Ever since, I remained curious about the phenomenon of cyclone, which is generally understood as an extreme low pressure that rapidly forms rapidly at sea. It drives inordinately fast winds, which on making landfall causes much havoc on the coast and inland. But how the cyclone forms has largely remained a mystery. There are many theories. I too have one to propose.

It is quite understandable when the sun heats up a desert and so the air above, which then rises to create a low pressure there. That in turn causes the cooler air from the nearby regions to blow in. But how does a highly localised and intense low pressure rapidly form over a small area at sea to become a cyclone? If heat is behind the phenomena of winds on the earth's surface, without some rapid heating of a small area over the sea, the cyclone just cannot form. And there is no other source of heat but the sun. But how can the sun heat up a small area at sea and so rapidly?

Is someone holding up a magnifying glass up in the sky to focus its rays sharply onto that zone, much the same away we burnt holes in paper or even started a fire using one? Probably yes, while it may not be a magnifying glass, but something quite similar—a lens forming naturally in the sky by some yet unknown concentration of gases and moisture present in the air.

The mirages in a desert or on a hot highway too are due to natural lens forming. There is no reason why that cannot happen elsewhere in the atmosphere. After all, the ingredients are same everywhere, air and heat. Such a lens forming in the upper atmosphere can actually be studied, perhaps quite easily from space, with so many satellites looking down to the earth. The key may be to measure the sudden and sharp changes in the atmospheric refractivity in a particular region, which may indicate the lens forming in the sky. It may then be possible to predict the cyclone formation long before it actually happens!

Let us, for the time being, at least, until someone comes up with the necessary empirical evidence, assume that such a lens does form in the sky, which then focuses the sun's rays onto a small area at sea. Obviously, the area will heat up rapidly. The seawater there may literally come to boil. Soon the air above will heat up rapidly and rise only to draw away the heat to the cooler atmosphere. That will make room for cooler air from surrounding regions to rapidly blow in.

Because of the Coriolis Effect, the winds blowing into the area will not take straight path, like spokes of a bicycle wheel, but curve in. With winds from all around curving in, the air in that zone will begin to spin, like a top. As the air in that area spins rapidly, the pressure there drops just as rapidly. That is in accordance with the Bernoulli's Law—faster the air flow greater the drop in pressure.

The rapidly spinning air soon heats up and rises to draw away the heat from the area, along with huge amount of moisture, like a gigantic suction pump. That further intensifies the drop in pressure in that zone. Therefore, the intense low pressure at the centre or eye of the cyclone is an effect, not the cause.

The rapidly rising column of hot, wet spinning air reaches great heights, often several kilometres high up into the upper atmosphere, where it condenses to form clouds that disperse radially, like a spiral, because of the spin. That is the familiar satellite picture of a cyclone—a spiralling band of clouds around a dark dot, the eye of cyclone.

A rapidly spinning object, even if it is a spinning column of air, has two interesting properties. One is rigidity. The spinning object tends to remain rigid in space. For example, a spinning top remains standing up on its point is because of this property of rigidity. The property of rigidity makes the cyclone to remain standing at a particular place, for some time. That in turn draws away the heat from there through the moisture it sucks and disperses all that into the cooler upper atmosphere. The cyclone wind system is therefore designed to dissipate the heat from a small area on the sea surface that somehow heated up rapidly.

The other property of the rapidly spinning object is its peculiar motion horizontally, which requires only a small external push. When we gently touch a spinning top, it does not move in the direction of touch, but at right angle. It seems as though the touch moves through ninety degrees in the direction of the top's spin before acting on it. The property is known as precession. The cyclone behaves no differently. A small force, which may be in the form of minor variations in wind force one side or the other, is sufficient to move the cyclone, even though it may weigh billions of tonnes. The movement of the cyclone as a system due to precession is however very slow compared to the winds blowing within. Moreover, the tiny forces act randomly. That makes the cyclone's movement on the sea surface almost impossible to predict. It can be compared to a top spinning on an uneven floor that changes direction abruptly and randomly.

When this slow randomly moving, spinning mass of air reaches a cold region, that is, a region with much less heat content, it dissipates rapidly. The heat content of land is much lower than sea. Therefore, the cyclone on making landfall rapidly dissipates. Without heat to be drawn to the cooler upper atmosphere, there is hardly a need for the cyclonic system to continue in existence!

With that brief foray into the cyclone, we can move on the story of monsoon, which is the wettest, fastest and longest blowing seasonal wind system in the world.

CHAPTER 46

STORY OF THE MONSOON

The monsoon makes landfall at the southwest corner of Indian Subcontinent around early-June, every year, and without fail. After making landfall, the winds go on blowing over the subcontinent for next four months bringing widespread rain, finally disappearing into the Thar Desert.

It is amazingly punctual, only because the sun is behind it, providing energy to drive the winds. Sun works with clockwork precision, never failing to stick to the routine. The sun's heat output however can wax and wane due to sunspots and solar flares. That in turn bears directly on the amount of heat reaching earth. As a result, the speeds at which the winds blow, including the seasonal ones like the monsoon, would vary.

The wind speed is directly proportional to the thermal gradient. The pressure drop along the sea surface as the wind blows is also directly proportional to its speed. That in turn has direct bearing on the amount of moisture the wind can suck from sea. Warmer the air wetter it can get. Conversely, colder the air, drier it is. The amount of moisture winds can bear therefore directly corresponds to the amount of rainfall they can deliver on making landfall.

Therefore, sun's heat output would have a much greater impact on the rainfall that monsoon brings to the subcontinent than many other factors currently being considered to predict the rainfall. Correlation of rainfall with solar activity therefore may provide a better prediction model for the monsoon rainfall.

For a wind system as massive and consistent as the monsoon, there must be a huge, stable heat source to start with. Since its destination is the northern hemisphere, the heat source to kick-start the winds must be in the southern hemisphere, necessarily along Tropic of Capricorn. We can therefore infer that monsoon starts out in December from the Australian Desert, when the sun is blazing down.

Some experts are of the view that the monsoon starts out closer home, around the Mascarene Islands, Reunion and Mauritius, where a high pressure zone develops around April every year. It has come to be called the Mascarene High. It is common belief that the wind blows from a high pressure. But monsoon starting out from the Mascarene High cannot account for the amount of rainfall it delivers to the subcontinent. Distance from the Mascarene High to the subcontinent is too short for the winds to pick up so much moisture from the sea. In addition, that makes it quite difficult to explain the heavy rains that must have fallen on the Indian Landmass to create the wide coastal continuum all around that we get to see today, particularly along the East Coast. This will become clear in the next chapter.

There are some who think that the monsoon starts more to the east, within the south equatorial belt, between 90 and 100 degrees east longitudes, that is, around Cocos and Christmas Islands. There too a high pressure forms, around January, every year. But for a high pressure to form, the air must sink in from above. That must necessarily come from a large heat source not far away. That is how I put the monsoon's start line at the Australian Desert, and in the month of December, when it is the hottest there.

As the Australian Desert heats up, air above heats up and rises. A low pressure forms over the desert, which draws in the cold air from south, which again heats up and rises. The process goes on. Rising air cools and heads north towards equator, where it is generally warmer. The Coriolis Effect deflects these high level winds to left, turning them into high-level south-easterlies.

As a convention, the direction of wind is reckoned as the direction from which it blows and not towards, which is the norm in every other case of reckoning direction of movement. Therefore, the south-easterlies are winds that blow from south-east towards north-west.

The south-easterlies then steadily sink into the warmer south equatorial belt around the Cocos and Christmas Islands. A high pressure soon develops there. From this high pressure, winds head out as easterlies, that is, winds blowing westward along the south equatorial belt.

As the easterlies blow through the south equatorial belt, there again is a significant drop in pressure along the belt, which besides causing moisture to be sucked from sea, also draws more winds from south. The winds from south blow into the belt as south-easterlies, because of the Coriolis Effect. The easterlies thus charged with more air and moisture blow unimpeded over the landless ocean towards Africa.

The winds then should have blown right into Africa, which by then had heated up due to the northerly shift of sun's declination. But that does not happen, because of the timely formation of the Mascarene High. Why this Mascarene High forms there, precisely at that time, is still a mystery, at least, to me. But the experts have confirmed its formation. I shall therefore leave it as a mystery and explore the end it serves to the monsoon phenomenon. Nature, we know, does nothing without a good purpose.

The winds cannot blow into to the high pressure, only away of it. The Mascarene High therefore deflects the moisture-laden easterlies northward, in other words, turns the easterlies into southerlies. That is also when Asian Landmass slowly begins to heat up to beckon winds into northern hemisphere, across the equator. On crossing the equator, the winds however do not remain as southerlies. Due to the Coriolis Effect, the southerlies are deflected right to become the south-westerlies, the Southwest Monsoon or the monsoon that blow right into the Indian Subcontinent, every year, at the appointed time.

What if the Indian Subcontinent was not there where it is today, as a part of Asia? That means the Himalayas would not have been there too. That would put Tibet more to south, lower and flatter. The southerly winds deflected into northern hemisphere by the Mascarene High, on turning right to become the south-westerlies, would have blown straight into Tibet bringing good rains, and so must have been a fecund landscape, unlike the cold rugged desert in the lee of Himalayas. In other words, the monsoon would have kept its schedule even without the Indian Subcontinent being there! But today, if there is a place most affected by the monsoon, it is the subcontinent. Life on the subcontinent depends on it. That is why it has come to be called the Indian Monsoon, better still, India's Monsoon.

While the subcontinent need not be there, where it is today, the monsoon phenomenon cannot do without the landmasses of Australia, Africa and Asia in their present positions or nearly so. Asia was more or less in its present position since long. Both Australia and Africa too did not take long to slip into their present positions after the breakup of Gondwanaland. These were nearly in their positions, unlike the Indian Landmass that had to journey a long way. Therefore, monsoon must have started to blow around 140 million years ago. It is an ancient phenomenon, certainly more ancient than the Indian Subcontinent that it serves today.

In making these deductions, I have made one assumption—the earth's axis of rotation remaining steady and at the same inclination as it is today, that is, 23 ½ degrees, since breakup of Gondwanaland. Only then will the Australian Desert heat up as it now does, in December. The axis of rotation however need not remain steady. It can tilt differently too.

Earth's mass is unevenly distributed around its axis of rotation, which makes the earth wobble as it rotates. Earth is rotating amazingly fast too. If you are somewhere on the equator, you are travelling at a speed close to 1,700 kmph. Spinning at that speed, it behaves like a gyro.

A slight wobble may then be sufficient to send it drifting into distant space due to the property of gyroscopic precession. That however does not happen because of the reining in by the sun's gravitational pull. But the wobble could change the tilt of axis, particularly when it gets too jerky.

It may have been quite jerky, when the earth's single landmass, Pangaea, split into Laurasia and Gondwanaland, some 200 million years ago, and later when these supercontinents further broke up not long after, barely another 35 million years later. I suspect the earth's axis must have tilted differently before each of these major continental breakups.

Steady tilt of the axis is however vital to the climate, as it exists today on the earth. If the axis tilted differently, the deserts would have been elsewhere. The seasonal winds, like monsoon, would therefore blow differently too.

With that story of monsoon, let us now examine its impact on the Indian Landmass, as it drifted from the southern to northern hemisphere, after breaking away from Gondwanaland. That is the story of the origin and evolution of Indian Coast, which from one end to the other, is the handiwork of monsoon, its winds, waves, rains and runoff—the Monsoon Coast!

CHAPTER 47

BRIEF HISTORY OF THE MONSOON COAST

We have gone through some happenings on the Indian Coast, only to highlight various aspects of the working of coast. Let us now see how the coast had its origin and evolved thereafter to the form we see today, as shown in the Figure 47.1 below.

On the map of Asia, the Indian Coast does appear to be a seamless part of the Asian Coast. But it is not. It is discernibly distinct. The subcontinent merged with Asia only recently, that is, just about 25 million years ago. Now that is not a long time ago in the earth's geological history. If we fit the earth's age, which is about 4,500 million years, on a day-long scale, the merger happened only about eight minutes ago!

About 165 million years ago, the subcontinent was only a tiny triangular landmass wedged into the northern side of the supercontinent Gondwanaland, which then lay far down south in the southern hemisphere. The Tethys Sea was to the north of the triangular Indian Landmass, if we could call it Indian yet. That was where the original Indian Coast actually was.

Along that original coast, to the south, stood a long and high mountain range, which today after millions of years of erosion has become motley hills and plateaus, stretching from the Aravalli Hills on its western end to Chotanagpur Plateau on the east.

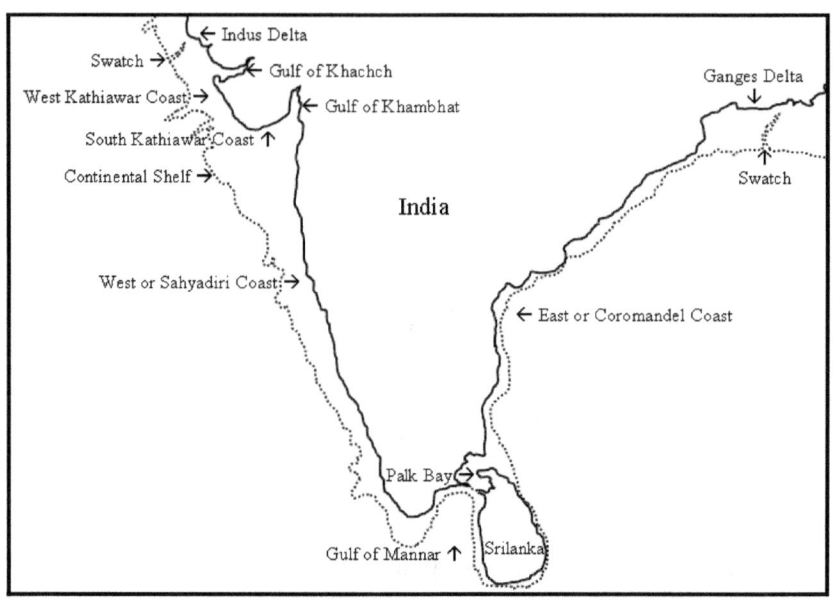

Figure 47.1: The Monsoon Coast

To the west of the landmass was Africa with the island of Madagascar sandwiched between. Along the entire length of the western margin, ran yet another long, high and unbroken mountain range, which was attached to the northern mountain range.

During my visit to Antarctica, in 1999, I collected several rock samples from the Schirmacher Oasis that stands on an elevated plateau. That is where 'Maitri', the Permanent Indian Station at Antarctica, set up. Incidentally, unlike other places, where we usually find snow and ice on high mountains, at Antarctica, the mountain tops and adjoining plateaus are usually ice-free. That was why the station was set up on the plateau. A station on ice shelf would steadily sink as ice beneath melts, because of the heat passed on by the station. India's first Antarctica station, Dakshin Gangotri, was set up on the ice shelf. In 1999, the station was more than ten metres below the surface of the ice-shelf, in a deep cavern of ice.

Back home, when I showed the rock samples to some of my friends, they were not particularly impressed, because they had seen plenty of similar rocks around. That had to be, because the part of Antarctica, where I collected the samples was once attached to the Indian Landmass along its entire eastern edge! Here too was a long, unbroken mountain range straddling the two landmasses. This mountain range spanned the entire length of the eastern margin, which at the north was attached to the eastern end of the northern mountain range, and in the south, connected to western mountain range at what has come to be called the Central Highlands of Srilanka.

So the triangular landmass had high unbroken mountain ranges on all its three sides. Hemmed within was a triangular plateau, the precursor to Deccan Plateau. The plateau was then a bone-dry desert, probably since millions of years, because no moisture could have got in, being hemmed by high mountains. The plateau also stood about three to five kilometres higher than it is today. That is also not a difficult inference.

On the Deccan Plateau are today numerous granite domes, solid mounds of granite, most of which are in the process of being quarried to meet the burgeoning demands of the construction industry, world over. From the granite's fine crystalline structure, we can infer that the magma had cooled rather slowly. That could not have happened on the surface where the granite domes are today, but only deep down, between three to five kilometres below the surface. That means all material above had been eroded away. That could not have happened without heavy rains falling, and for a very long time, for millions of years. So the plateau did receive heavy rains, eventually.

About 165 million years ago Gondwanaland began to breakup into smaller landmasses, continents of Africa, Antarctica and Australia and small triangular island, the Indian Landmass. After separating from the rest, the island Indian Landmass began to drift north. It covered more than 7,000 km to reach its present position about 140 million years later, moving about as fast as the human finger nail grows.

During the breakup, the mountain range on the west was sliced into two almost along it entire length. Foothills went off with Madagascar. The high range stayed on with Indian Landmass, which later eroded into hills of Western Ghats or Sahyadiri. A short while after separation of Madagascar, another sliver of land broke away from the landmass's western margin, the Mascarene Plateau, with Seychelles Islands to north, and to the south the islands of Mauritius and Reunion.

Before the breakup, Seychelles Islands were probably attached to the north mountain range, with Khachch and Kathiawar between. As Mascarene Plateau pulled away, Khachch and Kathiawar too were pulled apart from the landmass. But for some reason, the two remained nearby as islands. The two were in fact islands until as recent as 5,000 years ago, when the Indus Valley Civilization flourished. By then however the intervening sea had filled up with eroded sediments into a narrow waterway. When this waterway eventually filled up, Khachch and Kathiawar became integral parts of the Indian Subcontinent. That also brought curtains down on the civilisation.

In the south, most probably, Mauritius and Reunion were wedged within the hollow between India and Srilanka, where the Gulf of Mannar is today. The islands therefore could be the missing links between the Western Ghats and Central Highlands of Srilanka.

A friend, an avid birdwatcher, once told me about the parrots in Mauritius, which bore striking resemblance to those in southern Western Ghats, in spite of the huge geographical separation.

I told him, "Perhaps, someone must have carried a few parrots from India."

He replied, "That is not so. Parrots are endemic."

I told to him about how Mauritius was once a part of the Western Ghats and broke away along with Mascarene Plateau.

He asked, "How long ago did that happen?"

I replied, "About 150 million years ago."

Then he said, "Then it cannot be, because the parrots in Mauritius are of recent evolution."

I was stumped. I pondered. If parrots evolved from a common ancestor that existed before the breakup, and after breakup, conditions and so food habits remaining the same, more or less, at the southern Western Ghats and Mauritius, then there is every possibility that evolution must have produced similar outcomes, in spite of huge geographical separation. There are examples of evolutionary divergence after a species migrated elsewhere due to 'natural selection'. In the case of these parrots, it may not have been a case of the species migration, say, driven by an unseasonal storm, but their habitat changing location so slowly that change may have been hardly felt. It may warrant more research by evolutionary biologists to solve my friend's riddle.

On the east, the breakup of Antarctica did not go off quite as neatly. The mountain range did not get sliced along its length. Only a small part, a fifth, stayed back with the Indian Landmass. That is the 400 km long Mahendragiri Range. Rest went off with Antarctica. On Antarctica, the corresponding portion is today known as Queen Dronning Maud Land. Along the coast of Queen Dronning Maud Land runs a long mountain range. Napier Mountains are at its eastern end. Before the breakup, Napier Mountains were attached to the northern mountain range at around the Chotanagpur Plateau. The stretch corresponding to Mahendragiri Range is Prince Olav Coast, a flat coastal plain, covered by an ice-shelf. To the west of Prince Olav Coast is a long chain of mountains—Sor Rondane, Wohlthat and Muhlig-Holfmann Mountains and Ritscher Upland. These mountains were attached to the eastern margin to the south of Mahendragiri Range, reaching right down to Srilankan Central Highlands.

After the breakup, both western and eastern margins were steep, sheer vertical drop all the way down to seabed. In other words, there was no coastal continuum either side. The evolution of the continuum had to wait another 100 million years, because little rain fell on the landmass so long it remained far down south in the southern hemisphere. The rain bearing easterlies or the monsoon early on its journey to the northern hemisphere blew within the south equatorial belt.

As the landmass drifted north, the bed of Tethys Sea began to buckle. It was probably about 65 million years ago, when the landmass drifted into the south equatorial belt, did the seabed rise above the sea level, to eventually become the Himalayas!

As the Himalayas rose to majestic height, the island landmass merged with Asia to become a subcontinent. But the merger has not stopped the subcontinent's northerly drift. It continues even today, steadily pushing the Himalayas higher, albeit very slowly.

The zone of merger between Indian and Asian landmasses was initially a wide and deep canyon that stretched overland from one end of the coast to the other, with the new Himalayas to the north and the ancient north mountain range to south. But there is no canyon today. It has been filled up by sediments eroded from either mountain range, turning it into a vast plain, the Indo-Gangetic Plain, through which flow the two great river systems, Indus to the southwest and Ganges to east.

We can however see the canyon's seaward ends offshore of the two deltas, Indus and Ganges. These canyon-ends are known as 'swatches of no ground' or just swatches. See the Figure 47.1 above. Swatches are deep submarine valleys, with beds sloping steeply seaward. Western or Indus swatch is about 120 km long. It extends southwest from the delta to the bottom of Arabian Sea. Eastern or Ganges swatch extends about 300 km south to the bottom of Bay of Bengal.

If we extend the swatches coastward, along their lay, we can get to the ends of Indian Coast. But on the coast, there is nothing to show. Likewise, there is nothing on the coast to suggest the existence of swatches a short distance offshore. The swatches were discovered accidentally during the hydrographic surveys off the deltas, when the sounding machines, used those days to measure depths, failed to find bottom. It seemed as though there was no ground below, hence the name 'swatch of no ground'.

Swatches are transient features. In time, though not before millions of years, the sediments discharged by the rivers will obliterate them. But for the present, the two swatches make the Indian Coast quite distinct from rest of the Asian Coast.

Let us now examine how this distinctly unique coast evolved into the form we see today from the sheer vertical continental margin that it was at the time of breakup from Gondwanaland.

As the Indian Landmass crawled into south equatorial belt, the easterlies engaged it head on to the eastern margin. The desert plateau started to get heavy rains, probably the first heavy and sustained downpours since the breakup nearly 100 million years ago. Sediment-laden runoff started to flow slowly across the steep continental margin to the sea. The eastern coastal plain started to form as the margin accreted, with clay sticking to it, layer by layer. After a long time, the accreting margin went underwater and the continental shelf began to form. Thus, along most of the East Coast, the coastal plain is generally wide, whereas the continental shelf stayed narrow, with a steep drop seaward.

The easterlies first met the northeast corner of the landmass, around the present day Chotanagpur Plateau. Rains started there. Rivers also got flowing first there. Most probably, River Damodar was first big river to get flowing on the Indian Landmass. As it inched north into easterlies, more of the plateau began to get rains. Next big river to get flowing was the Mahanadi.

South of Mahanadi Basin, the easterlies encountered the then high Mahendragiri Range. As rains lashed the mountain side, sediments rushed to sea bottom to form continental shelf first, which eventually rose to surface to form the narrow coastal plain along the mountain.

As the landmass continued to drift north, rains spread across rest of the plateau, right up to the high mountain range along the western margin. The other peninsular rivers got flowing. Before long, as the plateau eroded away, the granite domes were exposed.

After the landmass moved further north, the easterlies began to blow through the gap south of western mountain range. The gap was formed when Mauritius and Reunion went off with the Mascarene Plateau long ago. Because the winds blew through, unstopped, little rains fell over the stretch. Therefore, very little sediment-laden runoff flowed across the eastern margin across that stretch. Intense waves however continued to lash the margin.

Without sediment supply, instead of accreting like the margin north, it began to erode. Erosion steadily progressed inland to create an open, but shallow bay, the Palk Bay. Erosion continued within the bay till the waves reached the natural rock-ledge, the Adam's Bridge or Rama Sethu. The erosion stopped there. The Adam's Bridge was just about at sea level. Thus

Srilanka became an island.

During the last ice-age, around 20,000 years ago, when the sea level dropped to lowest, Adam's Bridge must have shown up above sea level, like a land bridge to Srilanka, with narrow channels coursing through. If the legendary King of Ayodhya, Sri Rama, did go across to Srilanka, he could only have done so across this natural rock bridge by dumping mud and rocks to ford the channels within.

The Palk Bay then was a wide open bay. It narrowed into the shallow Palk Strait millions of years later. I shall come to that soon.

Before long, the landmass went on to straddle the south equatorial belt. That put an end to the flow of easterlies towards the Mascarene Plateau, thence to Tibet. That probably was the longest break that monsoon took since it got blowing, that too, stopped by the Indian Landmass! That caused the longest drought in Tibet, which never really ended. The transit of the landmass through south equatorial belt took about twenty million years. During this period, the East Coast evolved more or less to the form we see today, including the deltas of the major peninsular rivers. But there were no spits yet on this coast, because the waves from the easterlies remained head on to the coast. Spits came later.

During this period of intense rain, the once desert-like plateau turned into lush green forests. From these forests the runoff flowing through the major peninsular rivers delivered huge loads of biomass offshore across their deltas. That explains the gas finds off the deltas, but the reserves are unlikely to be spectacular, as in the case of submerged forests.

As the landmass transited through the south equatorial belt, its western margin stayed dry, steep and rugged, almost the way it was at the time of breakup. As the landmass began to move across the equator, the rains did not start off immediately, because it was still straddling south equatorial belt, cutting off the easterlies. Only after it moved up sufficiently north of equator, could the easterlies resume along the original track, to be deflected north by the Mascarene High and then blow as Southwest Monsoon, as before.

Because of the longer sea transit, the winds were lot wetter than the former easterlies, when they engaged the western margin. It may be difficult to imagine the intensity of rains that these wetter winds brought down, when abruptly stopped by a very high mountain range. Ensuing runoff may have been of deluge like proportion. The 250 cubic kilometres of runoff that flows annually down the Western Ghats today may only be a trickle in comparison!

Runoff rushing down the steep slopes brought down the mountain side in huge chunks, rocks and boulders hurtling down to the seabed. This initial deluge of mud and rocks deposited rapidly on the seabed probably formed the Chagos-Lakshadweep Plateau. This plateau does not appear to have been formed by continental breakup, like the Mascarene Plateau. Indian Landmass just deposited a part of its western margin on the seabed, because of intense rains and surging runoff, enough to rise up a plateau, and then drifted north.

Monsoon rains have been lashing this steep margin ever since. Rains and runoff razed the high mountain range by as much as four-fifth, depositing all that material on the seabed along the foot of the margin to form the wide continental shelf, which eventually rose above sea level to form a narrow coastal plain. There is an interesting legend regarding this rising of land from the sea.

Parasurama, the mace-bearing avatar of Vishnu, having killed his mother in a fit of rage, tried to expiate his sins by killing the Kshatriyas, who were then considered the enemies of Brahmins, and then by giving away all their land to the Brahmins. He was nevertheless disowned by the Brahmins and banished from the land.

So he was left with no option, but turn to the sea for a place to live. He went to Varuna, god of the Western Sea, today's Arabian Sea, seeking some land. But there was no level land between the mountain and sea. There was however enough level land along the Eastern Sea or Bay of Bengal. But that was the abode of Lakshmi, a goddess. He probably did not want to seek a favour from a goddess or probably the Brahmins were already living there.

Varuna offered Parasurama to gain from sea as much land as he could span in one throw of his mace. The gesture was not out of pity, but a ploy to rid Parasurama of his deadly mace. Left with little choice, standing at the edge of Western Mountain, Parasurama threw his mace into sea. Sea retreated. Coastal plain of Karnataka and

Kerala rose from the sea. At Gokarn Beach, south of Karwar, where the mace was believed to have fallen, stands a famous temple. It is now an important centre of pilgrimage for the Hindus.[11]

There is always some truth in every legend. The narrow coastal plain along the entire length of Western Ghats, not merely of Karnataka and Kerala, did rise from sea, though not by throw of mace.

The islands of Khachch and Kathiawar and the Aravalli Range received the first rains. Kathiawar went on to receive heavy rains for a long time, until it moved out of the scope of monsoon, just around the time of merger with Asia. That explains the wide continental shelf across the coast to west and south, off the West and South Kathiawar Coasts.

As the landmass inched north, rains began to lash the western mountain range, progressively. As a result, the sediment delivery to seabed by the runoff also happened progressively. The continental shelf therefore tapers down in width southward along this margin.

Width of coastal plain that rose from sea also follows a pattern. Where the adjoining mountains were high, the corresponding coastal plains were wide, for example, at Mangalore and Karwar. It was only a matter of sediment supply.

The western mountain range was unbroken along its entire length, but for the twenty kilometres wide Palakkad Gap. Offshore of this gap, the continental shelf takes a conspicuous detour landward due to relatively less sediment supply.

Next gap along this mountain range is Gulf of Mannar. Like the former easterlies, even the monsoon winds just blew through, bringing little rain to Mannar Coast between Kanyakumari and Adam's Bridge. As a result, the shelf along the coast stayed narrow, with steep seaward edge.

The southernmost tip of the subcontinent, Kanyakumari, extended more south. Intense monsoon waves, without adequate sediment supply,

---

[11] A slightly different version is in the Editorial Introduction: *Sahyadri* and Geological Time in Ancient Scripture—Gunnel Y. And Radhakrishna B. P. (Editors), *Memoir 47(1) Sahyadri The Great Escarpment of the Indian Subcontinent*, Bangalore: Geological Society of India, 2001. My version is based on oral traditions, what I heard from various folktales.

eroded Kanyakumari Coast to the present form, with indented rocky shoreline. The rocks off the Kanyakumari Coast, where some famous monuments stand, were once the part of a coastal landscape that stretched far into sea.

After the landmass drifted fully into the northern hemisphere, because of the high western mountain range blocking the monsoon, almost entirely, Deccan Plateau got no rains. It probably turned again into a desert or nearly so, with all its rivers drying up. It was probably several millions of years later, after the mountain range had eroded considerably, rains resumed and the rivers got flowing again.

As the landmass neared its present position, the heat zone, the Tropic of Cancer, came to be over Thar, which until then must have been a wet and fecund region, with several rivers and streams. Thar then turned into a desert.

Rest of the region along the Tropic of Cancer stayed fecund, because of the rivers systems, Indus and Ganges, got flowing. These rivers rise from the Himalayan glaciers. Had the Himalayas not risen above the 'snow line', most of the Indo-Gangetic Plain may have been a desert like the Thar. The plains could rapidly turn into a desert again, if the Himalayan glaciers melt away, say, due to some rapid global warming. So we must view global warming with greater concern, particularly on its impact on Himalayan glaciers.

As the Indian Landmass merged with Asia, the monsoon winds began to blow over to Bay of Bengal south of Srilanka. Drawn by the heat over the subcontinent, the winds turned anti-clockwise to blow, initially, as southerly winds along the East Coast, thereafter as south-easterlies across the rest of the subcontinent, to eventually end up at the Thar Desert as easterlies. That is where monsoon meets its end. That is the situation today.

Southerly winds did not bring much rain to the East Coast, but the waves that these winds generated met the coast at a broad angle from the south, setting off the longshore drift from south to north. That led to the formation of spits across the river mouths and bays. The permanent spits we see today were formed at that time. The first one to form was Jaffna

Spit, now known as Jaffna Peninsula, across the mouth of Palk Bay, narrowing it into the shallow Palk Strait. Longshore drift from Jaffna Spit delivered sediments into the strait rendering it shallow compared to the bay to west. Godavari Sand Spit across the Kakinada Bay too was formed around this time.

When the Indian Landmass merged with Asia, Kathiawar Peninsula moved out of the scope of monsoon rains. West Kathiawar Coast began to erode, but did not turn entirely rocky, because of the occasional sediment supply and milder waves from the westerlies. Since waves were at an angle to the coast, southerly longshore drift started. That in these recent times has been disrupted by several new ports and fishing harbours. As a result, the coast has begun to erode.

The monsoon waves however continued to lash the South Kathiawar Coast, but without commensurate sediment supply. So it soon turned rocky, entirely. The waves on the north Konkan Coast, though head on, became less intense. As a result, the coast turned silty and clayey. Further south, both rains and waves continued, more or less, as before. But the changes in the overland drainage at many places cut off sediment supply to the coast. At such places, the coast soon turned rocky or became cliffy headlands. The stretches between the headlands however stayed sandy, because of adequate sediment supply.

The rocky shorelines and cliffy headlands are closer to each on the Konkan Coast to the south of Mumbai. But on Malabar Coast, the headlands are quite far apart, with long beaches, in between. These stayed stable until quite recently till the coastal roads came to be built. The coast along these roads thus became deprived of the sediment supply that nature had all along ensured. Erosion is rampant along the West Coast, wherever the coastal roads have been built.

East Coast too is eroding, again quite rampantly, but due to the ports and other features that interrupted the longshore drift, which include the seawalls and groynes erected to the counter the erosion.

That in a nutshell is the history of the Monsoon Coast.

EPILOGUE

It took a long time for nature to shape the Indian Coast or for that matter any other coast in the world that are there today. But we, the humans, can undo all that in a matter of decades, if not faster. That is the kind of engineering prowess that we wield today. Destruction of the coast has however been largely due to ignorance. That must change. That is the aim of telling this story.

I hope this story will slowly worm its way into your hearts and minds, particularly of those of you who deal with the coast—conceiving, designing, planning, financing and building coastal projects—ports and other infrastructure on or near the coast.

When we do understand the coast, we can design and build projects that will not wreck it. Also, many problems currently faced due ill-conceived, ill-designed and ill-located coastal projects can be remedied to some extent, if not entirely. Of course, there will be a price to pay. That will also require bold decisions by those in the business of governance. But such bold and correct decisions can come from only those who know the coast well enough.

It was after an arduously long journey that I stumbled on what I believe is the truth. Again, I do not claim that to be the whole truth, because, as Alfred North Whitehead suggests, all truths are half-truths. Thereafter, it took me lot more time and effort to present that truth through this story. Since many things I have said go against the established science, I had to be double check everything, often many times over.

You know that story now, hence the truth, I hope. The knowledge of truth, even half-truth, comes with responsibility, which in this case is to protect the coast, to preserve it for the coming generations. They deserve the coast as nature crafted—not seawalls, groynes and saltpans.

Also please do not accept anything that I have said just because I claim that to be the truth. There is a need to question everything, challenge everything, and investigate everything. There still remains much more to be known. In the bargain, you too will stumble on the truth that you can believe. Truth can only be believed. There is no other way of knowing truth.

The quest for truth is the primary, perhaps only purpose of our life on earth, both as individuals and as species. It is for that alone we have been so graciously endowed with the remarkable trait of curiosity that no other species possess. This story would have served its end if it has aroused your curiosity, not merely about the coast, but about everything around.

Curiosity holds the key to unravel the countless mysteries of nature that still remain beyond our reach. Most of us, for some reason or the other, lose this trait soon after the days of childhood. We turn cautious, instead of being curious. We stop questioning and choose only the known and familiar. We turn conformists, and so we accept blindly what has been handed down by the past or by those deemed as authorities, temporal or intellectual. In other words, we cease to learn. Curiosity we possessed as children alone can drive the learning process that seeks out new truths. Therefore, to keep learning, there is a need to remain as children, the real learning machines.

Keep therefore your curiosity ever alive. There are many more mysteries waiting to be unravelled, but only through the child in you. How I wish I could grant you this wish—*"May you be blessed with such an undying curiosity!"* That may seem more like a curse, when we consider the troubles we may have to endure in the quest for truth that curiosity will invariably drive. But when you do stumble on the truth, whenever, the reward will invariably outweigh every trouble you endured. Try me.

When Archimedes stumbled on the truth about distinguishing a genuine gold crown from a counterfeit one, he jumped out of his bathtub and ran naked through the streets shouting, "Eureka! Eureka!", meaning "I have found it! I have found it!" That, I believe, was too mild an expression of the joy of stumbling on truth.

Without curiosity many wonders slip past without our knowing. That is too great a loss indeed, when we consider our brief existence as living human beings on this planet.

Curiosity, as I have already mentioned in the beginning of this story, is the mother of science. That is why I have classed this book as a book of science. So here is what one of the greatest scientists, who walked Planet Earth, had to say about curiosity.

"The important thing is not to stop questioning. Curiosity has its own reason for existing. One cannot help but be in awe when he contemplates the mysteries of eternity, of life, of the marvellous structure of reality. It is enough if one tries merely to comprehend a little of this mystery every day. Never lose a holy curiosity."

—*Albert Einstein*

THE END

REFERENCES

—*Admiralty Manual of Hydrographic Surveying, Volume One*, London: Hydrographer of the Navy, 1965.

—*Admiralty Manual of Hydrographic Surveying, Volume Two*, London: Hydrographer of the Navy, 1970.

—*Antarctica*, Australia: Reader's Digest (Australia) Pty Ltd, 1990.

—*Atlas of the World*, London: Reader's Digest Association Limited, 1997.

—*Bay of Bengal Pilot, INP 2 (First Edition)*, Dehradun: Chief Hydrographer to the Government of India, 2007

Bill, Bryson, *A Short History of Nearly Everything*, New York: Broadway Books, 2003.

—*Breakers and Surf, Principles in Forecasting, HO Publication No. 234*, Washington DC: US Naval Oceanographic Office, 1969.

Burroughs, William (Editor), *Climate into the 21st Century – World Meteorological Organisation*, Cambridge: Cambridge University Press, 2003.

—*Coastal Engineering Manual (Parts I, II, III and IV)*, Washington DC: US Army Corps of engineers, 2002.

—*Coastal Geology*, Washington DC: US Army Corps of engineers, 1995.

—*Coastal Littoral Transport*, Washington DC: US Army Corps of engineers, 1992.

Downs, Peter W., and Gregory, Kenneth J., *River Channel Management Towards Sustainable Catchment Hydrosystems*, London: Arnold Publishers, 2003.

—*Dredging and Disposal of Dredged Material – Engineering and Design*, Washington DC: US Army Corps of engineers, 1983.

—*Dredging for Development*, Report of the Dredging Task Force (Third Revised Edition) to the 17th IAPH Conference Barcelona, Spain, 1991.

—*Fairy Tales of Hans Christian Andersen*, London: Reader's Digest Association Limited, 2009.

Frederick K. L. and Edward J. T, *Essentials of Geology (Ninth Edition)*, New Jersey: Pearson Prentice Hall, 2006.

—*Geophysical Exploration for Engineering and Environmental Investigations*, Washington DC: US Army Corps of engineers, 1995.

—*Green Rating Project, Environmental Rating of India Caustic-Chlorine Sector*, New Delhi: Centre for Science and Environment, 2002.

Gunnel Y. And Radhakrishna B. P. (Editors), *Memoir 47(1) Sahyadri The Great Escarpment of the Indian Subcontinent*, Bangalore: Geological Society of India, 2001.

Gunnel Y. And Radhakrishna B. P. (Editors), *Memoir 47(2) Sahyadri The Great Escarpment of the Indian Subcontinent*, Bangalore: Geological Society of India, 2001.

Hiranadani G. M., *Transition to Triumph History of Indian Navy 1965-1975*, New Delhi: Naval Headquarters, 1999.

Hiranadani G. M., *Transition to Eminence History of Indian Navy 1976-1990*, New Delhi: Naval Headquarters, 2004.

Hiranandani M. G. and Gole C. V., *Use of Radioactive Tracer for the Study of Sediment Movement Off Bombay Harbour, Technical Memorandum I*, Poona: Central Water and Power Research Station, 1960.

—*Hydrographic Dictionary, Part I, Volume 1 English, Special Publication No. 32(Fifth Edition)*, Monaco: International Hydrographic Organization, 1994.

—*Hydrographic Surveying*, Washington DC: US Army Corps of engineers, 2002.

—*Illustrated Atlas of the World*, London: Reader's Digest Association Limited, 1997.

—*India Port Report, Ten Years of Reforms and Challenges Ahead*, Mumbai: i-maritime Consultancy Private Limited, 2003.

—*Indian List of Lights & Fog Signals, Vol D & E, INP 21 (a)*, Dehradun: Chief Hydrographer to the Government of India, 2003.

—*Indian List of Lights & Fog Signals, Vol F & K, INP 21 (b)*, Dehradun: National Hydrographic Office, 2003.

Julien, Pierre Y., *Erosion and Sedimentation*, Cambridge: Cambridge University Press, 1998.

Levin H. L., *Contemporary Physical Geology (Third Edition)*, Philadelphia: Saunders College Publishing, 1990.

Lloyd, Julie, *A Concise Guide to Weather*, Bath: Parragon Books, 2007.

Lutgens F. K., and Tarbuck E. J., *Essentials of Geology (Ninth Edition)*, New Jersey: Pearson Education International, Prentice Hall, 2006.

—*Manual on Hydrography, Publication M-13 (First Edition)*, Monaco: International Hydrographic Organization, 2005.

McGregor, Glen R. and Nieuwolt, Simon, *Tropical Climatology (Second Edition)*, New Jersey: John Wiley & Sons, 1998.

Menachery G. and Chakkalakkal W., *Kodungallur The Cradle of Christianity in India*, Kodungallur: Mar Thoma Pontifical Shrine, 2000.

Narasimhan S., Kathiroli S., and Nagendra, Kumar B., *Harbour and Coastal Engineering [Indian Scenario] Volumes I & II*, Chennai: National Institute of Ocean Technology, 2002.

—*Oxford Dictionary of Quotations (Seventh Edition)*, Oxford: Oxford University Press, 2009.

Radhakrishna B. P., and Vaidyanathan R., *Geology of Karnataka*, Bangalore: Geological Society of India, 1997.

References

Richard, Overy and Geoffrey Barraclough (Editors), *Complete History of the World*, New Delhi: HarperCollins India, 2009.

—*Satellite Atlas of the World*, Washington DC: National Geographic Society, 1998.

—*Sethusamudram Ship Canal Project, DPR and Evaluation of EIA Study*, L&T Ramboll Consulting Engineers Limited, Chennai, 2005.

—*Sethusamudram Ship Canal Project, Full Version of Environmental Impact Assessment*, National Environmental Engineering Research Institute, Nagpur, 2004.

—*Shore Protection Manual*, Washington DC: US Army Corps of engineers, 1977.

Smith, David G., *The Cambridge Encyclopaedia of Earth Sciences*, Cambridge: Cambridge University Press, 1981.

Sridharan K., *A Maritime History of India*, New Delhi: Director Publication Division, Ministry of Information and Broadcasting, 1982.

Sridharan K., *History of Naval Dockyard Bombay*, New Delhi: Naval Headquarters, 1989.

Soman K., *Geology of Kerala*, Bangalore: Geological Society of India, 2002.

Sparks B. W., *Geomorphology (Third Edition)*, London: Longman, 1986.

—*The Mariner's Handbook, NP 100 (Fourth Edition)*, London: Hydrographer of the Navy, 1973.

Vaidyanathan R., and Ramakrishnan M., *Geology of India Volumes I & II*, Bangalore: Geological Society of India, 2008.

Walker, Gabrielle, *An Ocean of Air A Natural History of the Atmosphere*, London: Bloomsbury Publishing Plc., 2007.

—*Water Wave Mechanics*, Washington DC: US Army Corps of engineers, 2002.

—*West Coast of India Pilot, INP 1 (Second Edition)*, Dehradun: Chief Hydrographer to the Government of India, 2003

—*When Nature Turns Nasty*, London: Reader's Digest Association Limited, 2007.

INDEX

ABOUT THE AUTHOR

Commander John Jacob Puthur, Indian Navy (Retired)

Commander Puthur has devoted the best part of his life to study the Indian Coast. This book is the outcome. He is an alumnus of National Defence Academy, Khadakvasla, commissioned in the Indian Navy on 01 Jan 1980 in the Executive Branch. In 1982, he joined the Survey Branch and specialised in hydrographic surveying. He has served aboard several surveying ships in various capacities. From 1994 to 97 he was Instructor and later the Chief Instructor of National Institute of Hydrography, Goa. In 1998, he was graded 'Charge Hydrographer', the highest certification for a hydrographic surveyor accorded by the Chief Hydrographer to the Government of India. Also in 1998, he led a hydrographic surveying team to Antarctica, with the XVIII Indian Antarctica Expedition. Since premature retirement in September 2003, he has been a consultant in hydrographic surveying and port design. He is a Fellow of Indian Institution of Surveyors, Indian National Cartographic Association, and Geological Society of India. He is also a Life Member of Indian Society of Remote Sensing and Eastern Dredging Association (India). He lives in Bangalore with his wife and two daughters.

www.ingramcontent.com/pod-product-compliance
Lightning Source LLC
Chambersburg PA
CBHW072032190526
45165CB00017B/143